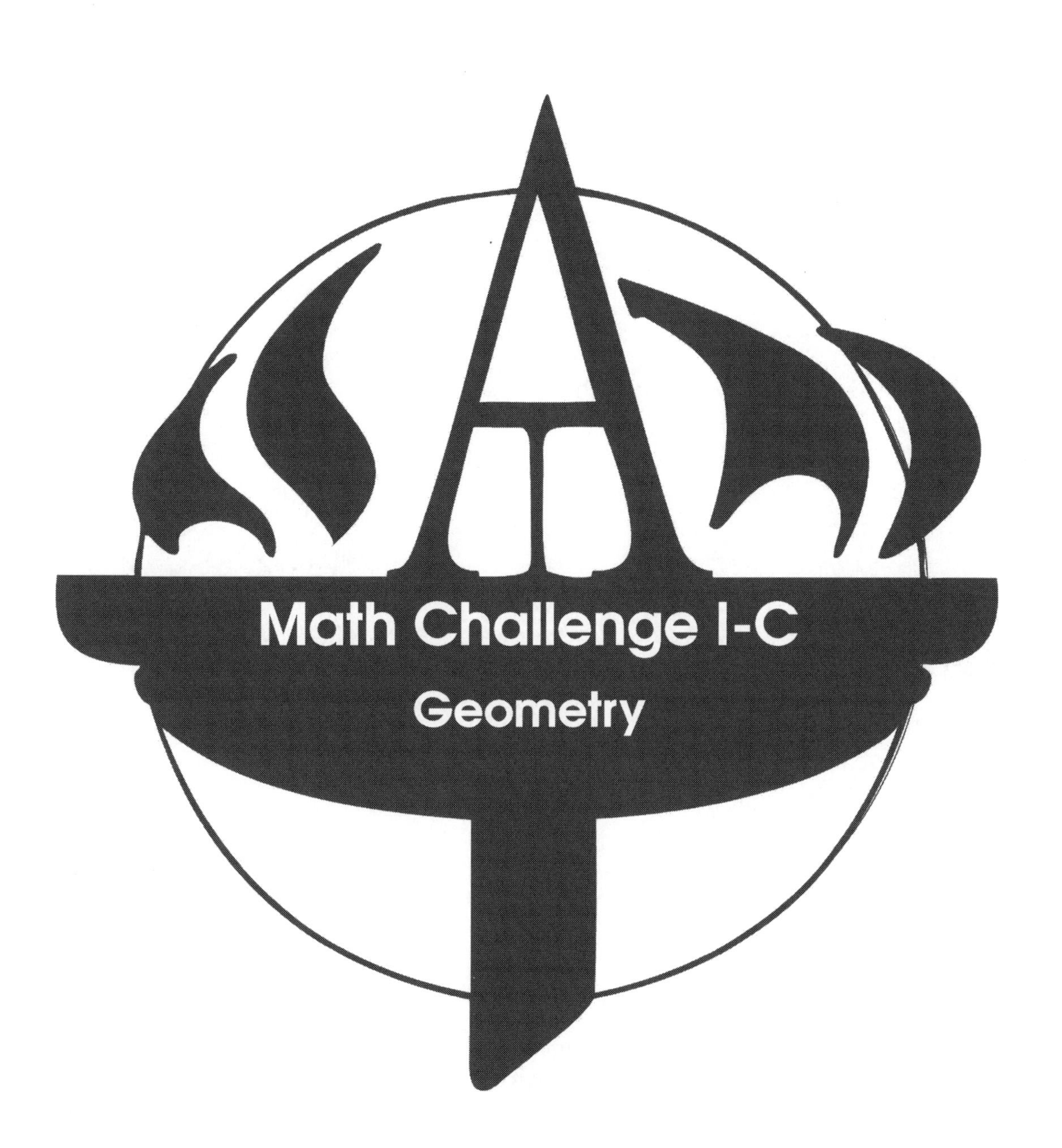

Areteem Institute

Math Challenge I-C Geometry

Edited by David Reynoso
John Lensmire
Kevin Wang
Kelly Ren

WWW.ARETEEM.ORG

PUBLISHED BY ARETEEM PRESS

ISBN: 1-944863-34-6
ISBN-13: 978-1-944863-34-0
First printing, November 2018.

TITLES PUBLISHED BY ARETEEM PRESS

Cracking the High School Math Competitions (and Solutions Manual) - Covering AMC 10 & 12, ARML, and ZIML
Mathematical Wisdom in Everyday Life (and Solutions Manual) - From Common Core to Math Competitions
Geometry Problem Solving for Middle School (and Solutions Manual) - From Common Core to Math Competitions
Fun Math Problem Solving For Elementary School (and Solutions Manual)

ZIML MATH COMPETITION BOOK SERIES

ZIML Math Competition Book Division E 2016-2017
ZIML Math Competition Book Division M 2016-2017
ZIML Math Competition Book Division H 2016-2017
ZIML Math Competition Book Jr Varsity 2016-2017
ZIML Math Competition Book Varsity Division 2016-2017
ZIML Math Competition Book Division E 2017-2018
ZIML Math Competition Book Division M 2017-2018
ZIML Math Competition Book Division H 2017-2018
ZIML Math Competition Book Jr Varsity 2017-2018
ZIML Math Competition Book Varsity Division 2017-2018

MATH CHALLENGE CURRICULUM TEXTBOOKS SERIES

Math Challenge I-A Pre-Algebra and Word Problems
Math Challenge I-B Pre-Algebra and Word Problems
Math Challenge I-C Algebra
Math Challenge II-A Algebra
Math Challenge II-B Algebra
Math Challenge III Algebra
Math Challenge I-A Geometry
Math Challenge I-B Geometry
Math Challenge I-C Topics in Algebra
Math Challenge II-A Geometry
Math Challenge II-B Geometry
Math Challenge III Geometry
Math Challenge I-A Counting and Probability
Math Challenge I-B Counting and Probability
Math Challenge I-C Geometry
Math Challenge II-A Combinatorics
Math Challenge II-B Combinatorics

Math Challenge III Combinatorics
Math Challenge I-B Number Theory
Math Challenge II-A Number Theory

COMING SOON FROM ARETEEM PRESS

Fun Math Problem Solving For Elementary School Vol. 2 (and Solutions Manual)
Counting & Probability for Middle School (and Solutions Manual) - From Common Core to Math Competitions
Number Theory Problem Solving for Middle School (and Solutions Manual) - From Common Core to Math Competitions
Other volumes in the **Math Challenge Curriculum Textbooks Series**

The books are available in paperback and eBook formats (including Kindle and other formats). To order the books, visit `https://areteem.org/bookstore`.

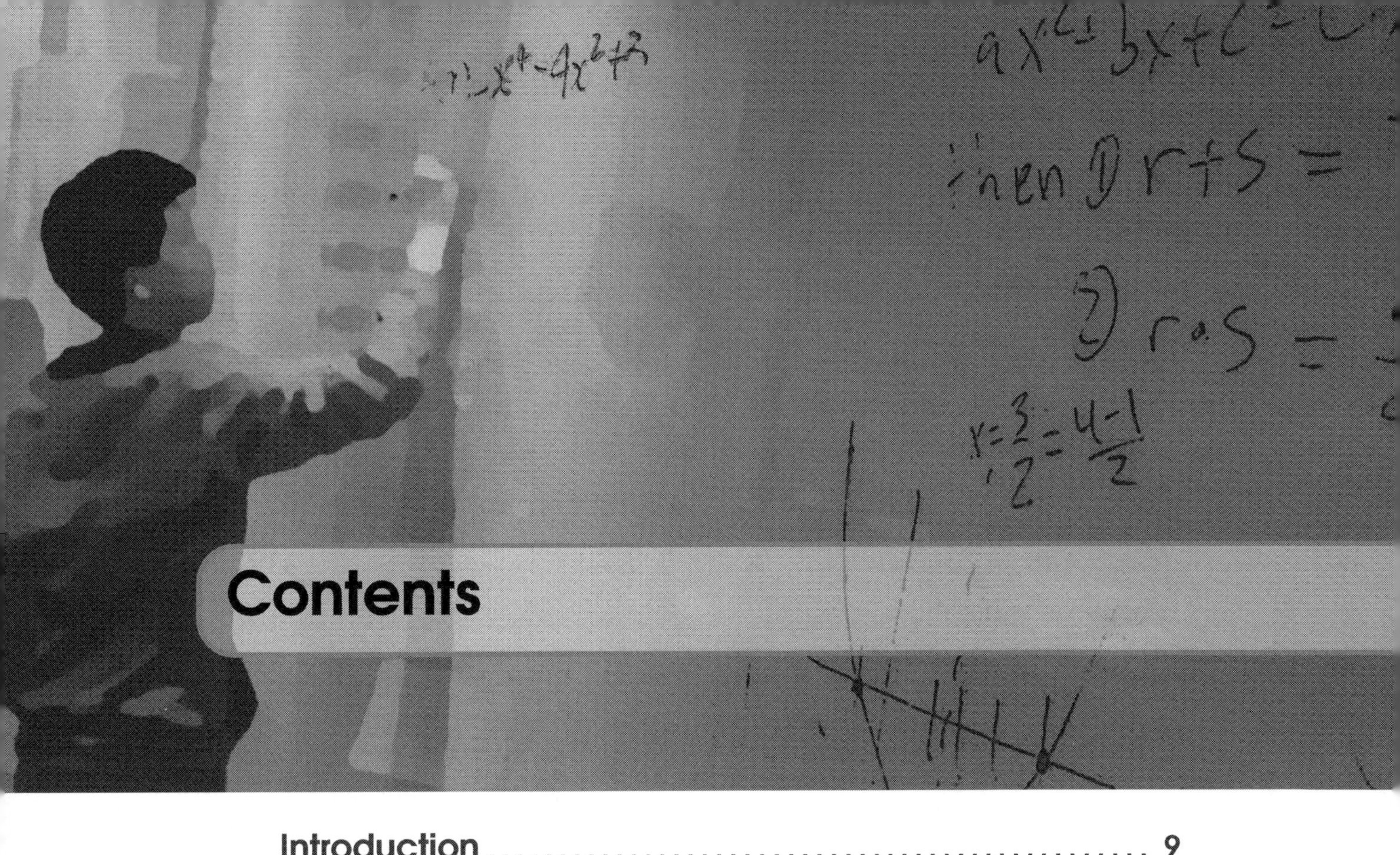

Contents

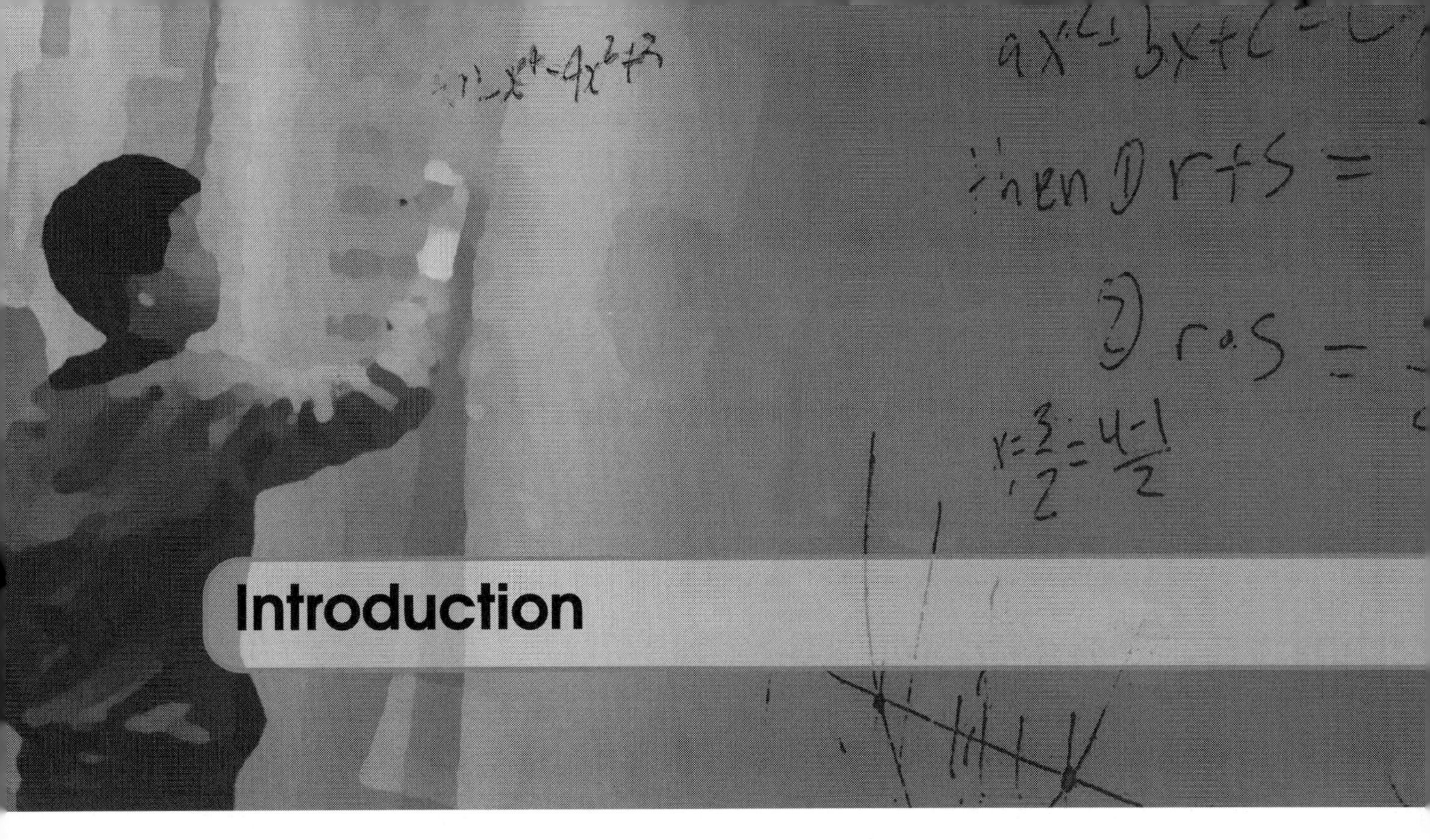

Introduction

The math challenge curriculum textbook series is designed to help students learn the fundamental mathematical concepts and practice their in-depth problem solving skills with selected exercise problems. Ideally, these textbooks are used together with Areteem Institute's corresponding courses, either taken as live classes or as self-paced classes. According to the experience levels of the students in mathematics, the following courses are offered:

- Fun Math Problem Solving for Elementary School (grades 3-5)
- Algebra Readiness (grade 5; preparing for middle school)
- Math Challenge I-A Series (grades 6-8; intro to problem solving)
- Math Challenge I-B Series (grades 6-8; intro to math contests e.g. AMC 8, ZIML Div M)
- Math Challenge I-C Series (grades 6-8; topics bridging middle and high schools)
- Math Challenge II-A Series (grades 9+ or younger students preparing for AMC 10)
- Math Challenge II-B Series (grades 9+ or younger students preparing for AMC 12)
- Math Challenge III Series (preparing for AIME, ZIML Varsity, or equivalent contests)
- Math Challenge IV Series (Math Olympiad level problem solving)

These courses are designed and developed by educational experts and industry professionals to bring real world applications into the STEM education. These programs are ideal for students who wish to win in Math Competitions (AMC, AIME, USAMO, IMO,

ARML, MathCounts, Math League, Math Olympiad, ZIML, etc.), Science Fairs (County Science Fairs, State Science Fairs, national programs like Intel Science and Engineering Fair, etc.) and Science Olympiad, or purely want to enrich their academic lives by taking more challenges and developing outstanding analytical, logical thinking and creative problem solving skills.

Math Challenge I-C is a four-part course designed to bridge the middle school and high school math materials. For students who participate in the American Math Competitions (AMC), there is a big gap in both the fundamental math concepts and the problem-solving techniques involved between the AMC 8 and AMC 10 contests. This course is developed to help students transition smoothly from middle school to high school, and prepare them for high school math competitions including the AMC 10 & 12, ARML, and ZIML. The full course covers topics and introductory problem solving in algebra, geometry, and finite math. Algebraic topics include linear equations, systems of equations and inequalities, exponents and radicals, factoring polynomials, and solving quadratic equations. Geometric topics include angles in triangles, quadrilaterals, and polygons, congruent and similar polygons, calculating area, and algebraic geometry. Topics in finite math include logic, introductory number theory, and an introduction to probability and statistics. These topics serve as the fundamental knowledge needed for a more advanced problem solving course such as Math Challenge II-A.

The course is divided into four terms:

- Summer, covering Algebra
- Fall, covering covering additional topics in Algebra
- Winter, covering Geometry
- Spring, covering Finite Math

The book contains course materials for Math Challenge I-C: Geometry.

We recommend that students take all four terms starting with the Summer, but students with the required background are welcome to join for later terms in the course.

Students can sign up for the live or self-paced course at `classes.areteem.org`.

About Areteem Institute

Areteem Institute is an educational institution that develops and provides in-depth and advanced math and science programs for K-12 (Elementary School, Middle School, and High School) students and teachers. Areteem programs are accredited supplementary programs by the Western Association of Schools and Colleges (WASC). Students may attend the Areteem Institute in one or more of the following options:

- Live and real-time face-to-face online classes with audio, video, interactive online whiteboard, and text chatting capabilities;
- Self-paced classes by watching the recordings of the live classes;
- Short video courses for trending math, science, technology, engineering, English, and social studies topics;
- Summer Intensive Camps held on prestigious university campuses and Winter Boot Camps;
- Practice with selected free daily problems and monthly ZIML competitions at ziml.areteem.org.

Areteem courses are designed and developed by educational experts and industry professionals to bring real world applications into STEM education. The programs are ideal for students who wish to build their mathematical strength in order to excel academically and eventually win in Math Competitions (AMC, AIME, USAMO, IMO, ARML, MathCounts, Math Olympiad, ZIML, and other math leagues and tournaments, etc.), Science Fairs (County Science Fairs, State Science Fairs, national programs like Intel Science and Engineering Fair, etc.) and Science Olympiads, or for students who purely want to enrich their academic lives by taking more challenging courses and developing outstanding analytical, logical, and creative problem solving skills.

Since 2004 Areteem Institute has been teaching with methodology that is highly promoted by the new Common Core State Standards: stressing the conceptual level understanding of the math concepts, problem solving techniques, and solving problems with real world applications. With the guidance from experienced and passionate professors, students are motivated to explore concepts deeper by identifying an interesting problem, researching it, analyzing it, and using a critical thinking approach to come up with multiple solutions.

Thousands of math students who have been trained at Areteem have achieved top honors and earned top awards in major national and international math competitions, including Gold Medalists in the International Math Olympiad (IMO), top winners and qualifiers at the USA Math Olympiad (USAMO/JMO) and AIME, top winners at the

Zoom International Math League (ZIML), and top winners at the MathCounts National Competition. Many Areteem Alumni have graduated from high school and gone on to enter their dream colleges such as MIT, Cal Tech, Harvard, Stanford, Yale, Princeton, U Penn, Harvey Mudd College, UC Berkeley, or UCLA. Those who have graduated from colleges are now playing important roles in their fields of endeavor.

Further information about Areteem Institute, as well as updates and errata of this book, can be found online at http://www.areteem.org.

Acknowledgments

This book contains many years of collaborative work by the staff of Areteem Institute. This book could not have existed without their efforts. Huge thanks go to the Areteem staff for their contributions!

The examples and problems in this book were either created by the Areteem staff or adapted from various sources, including other books and online resources. Especially, some good problems from previous math competitions and contests such as AMC, AIME, ARML, MATHCOUNTS, and ZIML are chosen as examples to illustrate concepts or problem-solving techniques. The original resources are credited whenever possible. However, it is not practical to list all such resources. We extend our gratitude to the original authors of all these resources.

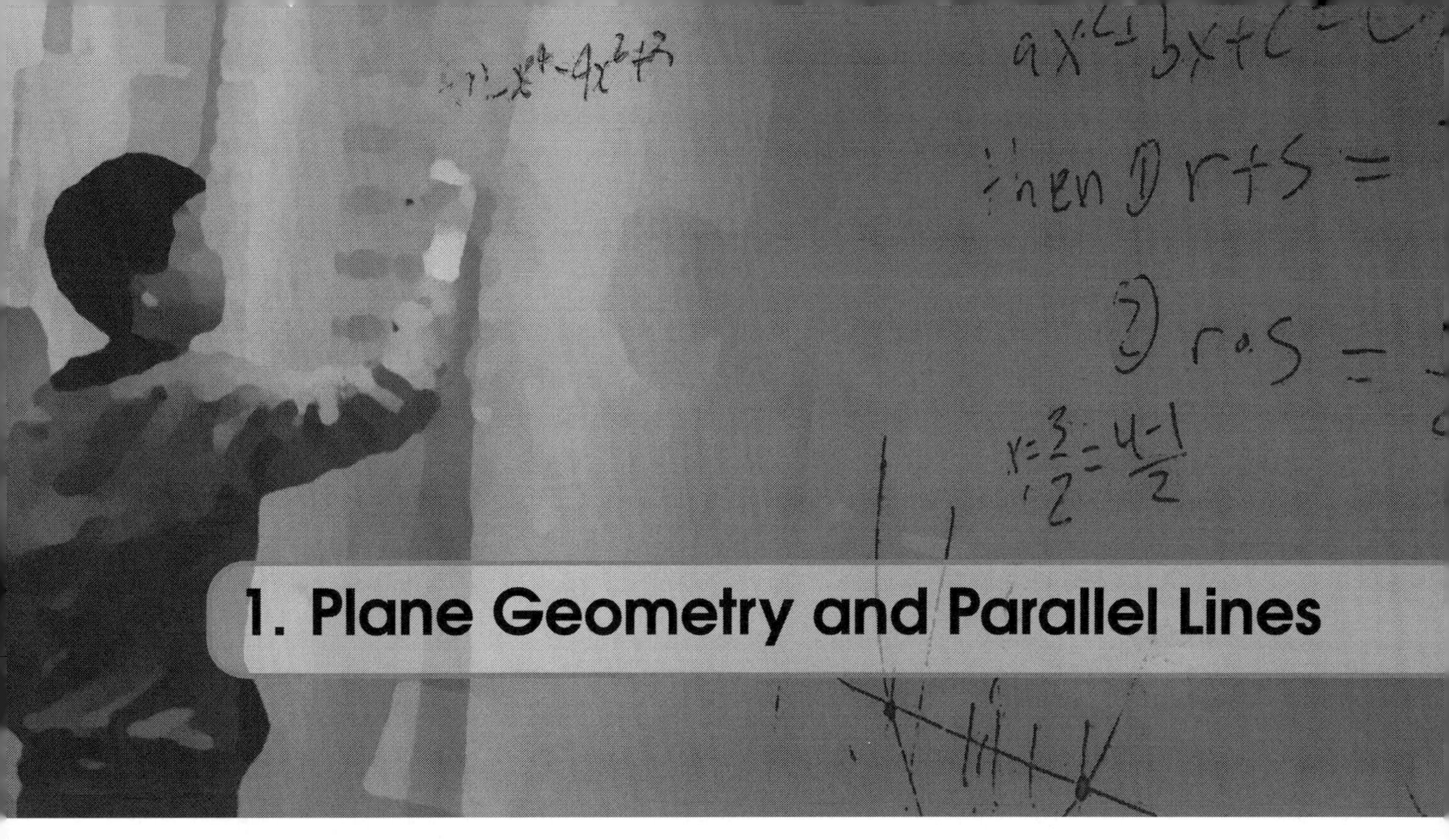

1. Plane Geometry and Parallel Lines

Plane Geometry and Coordinates

- A plane is a two-dimensional flat geometric object that extends forever. Think of a flat piece of paper with no boundaries.
- Plane geometry, also called, 2-D geometry, takes place inside a plane.
- Thus, it can sometimes be useful to label points in the coordinate plane, with the x-axis representing horizontal distance and the y-axis representing vertical distance as shown below.

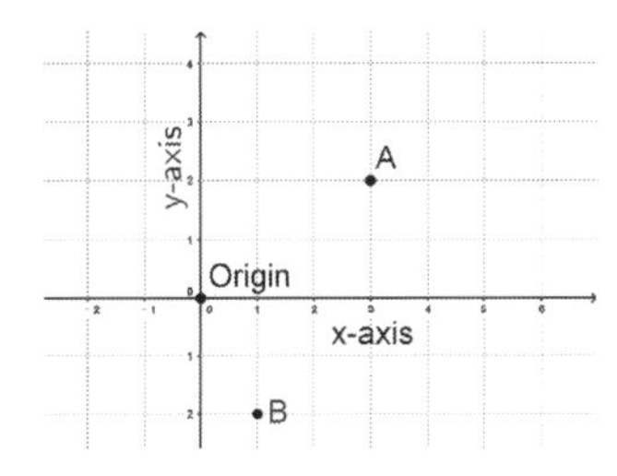

  - The point A above has coordinates $(x_A, y_A) = (3,2)$ and the point B above has coordinates $(x_B, y_B) = (1,-2)$.
- Example (The Midpoint Formula): The midpoint of $\overline{AB}$ is $\left(\dfrac{x_A+x_B}{2}, \dfrac{y_A+y_B}{2}\right)$.

Review of Slopes

- Slope between two points A and B: $\dfrac{y_A - y_B}{x_A - x_B}$.
- Horizontal Lines have slope 0.

- Vertical Lines have undefined or infinite slope.
- Parallel Lines have the same slope.
- Perpendicular Lines have opposite reciprocal slopes. That is, if you multiply the slopes together you get -1.

Review of Lines

- Standard form: $Ax+By=C$.
- Slope-intercept form: $y=mx+b$ (m is the slope, b is the y-intercept).
- Point-slope form: $y-y_A=m(x-x_A)$ for a line with slope m going through point A.

Angles

- Two lines that intersect form angles. We often measure angles in degrees.
- To classify angles:
  - An angle less than 90° is called an acute angle.
  - An angle equal to 90° is called a right angle.
  - An angle between 90° and 180° is called an obtuse angle.
  - An angle equal to 180° is called a straight angle.
  - An angle greater than 180° is called a reflex angle.
  - A full circle is 360°.

Angles in Intersecting Lines

- Consider the diagram below with angles $\angle 1, \angle 2, \angle 3, \angle 4$.

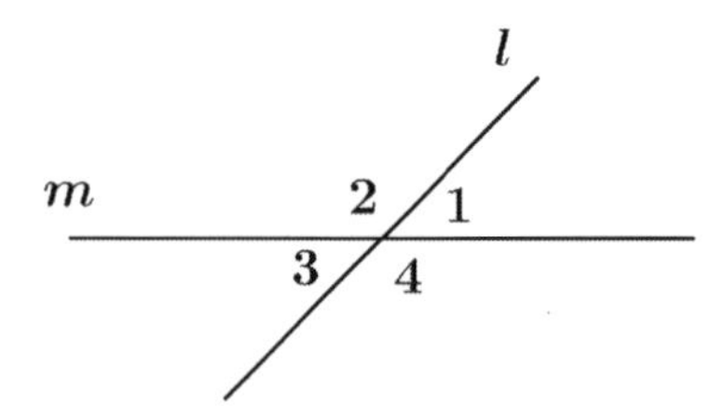

  - We call angles $\angle 1, \angle 3$ or $\angle 2, \angle 4$ vertical angles. Vertical angles are equal so $\angle 1=\angle 3$ and $\angle 2=\angle 4$.
  - We call angles such as $\angle 1, \angle 2$ adjacent angles. These angles add up to 180°, so $\angle 1+\angle 2=180^\circ$.
- In general, we call angles that add up to 180° supplementary angles.
- Similarly, we call angles that add up to 90° complementary angles.

Angles in Parallel Lines

- We use the notation $m \parallel n$ to denote the lines m, n are parallel. Line l intersecting both parallel lines is referred to as a transversal, as in the diagram below.

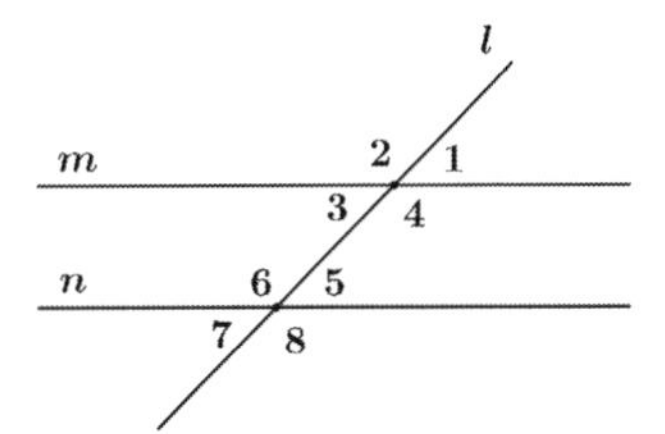

  - Angles such as $\angle 1, \angle 5$ or $\angle 4, \angle 8$ are called corresponding angles.
  - Angles such as $\angle 3, \angle 5$ or $\angle 4, \angle 6$ are called alternating interior angles.
  - Angles such as $\angle 1, \angle 7$ or $\angle 2, \angle 8$ are called alternating exterior angles.
  - Angles such as $\angle 3, \angle 6$ or $\angle 4, \angle 5$ are called same-side interior angles.
  - Angles such as $\angle 1, \angle 8$ or $\angle 2, \angle 7$ are called same-side exterior angles.
- Pairs of angles that are corresponding or alternating (interior or exterior) are always equal. Therefore, $\angle 1 = \angle 3 = \angle 5 = \angle 7$ and $\angle 2 = \angle 4 = \angle 6 = \angle 8$.
- Pairs of angles that are same-side (interior or exterior) are always supplementary.

1.1 Example Questions

Problem 1.1 Let $A = (0,0)$, $B = (-2,1)$, and $C = (3,6)$.

(a) Plot the points A, B, and C on a coordinate plane.

(b) Find the equation of the line containing points A and B.

(c) Find the equation of the line containing points B and C.

Problem 1.2 Prove that the equation for the line going through points A and B is given by

$$\frac{x-x_A}{x_A-x_B}=\frac{y-y_A}{y_A-y_B}.$$

Problem 1.3 Let $A=(1,2)$ and $B=(3,y_B)$. What is y_B if $\overline{AB}$ is parallel the line $y=3x-4$?

Problem 1.4 Given a line segment $\overline{AB}$, the perpendicular bisector of $\overline{AB}$ is a line perpendicular to $\overline{AB}$ going through its midpoint.

If $A=(1,1)$ and $B=(5,3)$, find the perpendicular bisector of $\overline{AB}$.

Problem 1.5 Let $O=(0,0)$, $A=(-1,0)$, and $C=(1,0)$. If B is a point such that $\angle AOB=30^\circ+\angle BOC$, what is $\angle AOB$?

Problem 1.6 Consider the diagram below, where l and m are parallel but the drawing is not necessarily to scale.

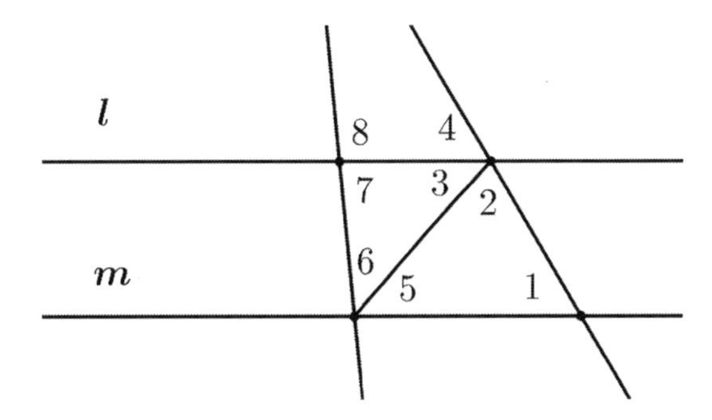

Suppose that $\angle 1 = 60^\circ, \angle 5 = 50^\circ, \angle 8 = 105^\circ$. Find the measure of $\angle 6$.

Problem 1.7 Prove that the angles in any triangle ABC add up to 180°.

Problem 1.8 Let $\angle A, \angle B, \angle C$ be the angles in triangle $\triangle ABC$. If $\angle B - \angle A = \angle C - \angle B$ and $\angle A = 33^\circ$, what is the angle measure of $\angle C$?

Problem 1.9 Consider the lines $y = 0$, $y = mx + 1$, and $y = \frac{-1}{m}x - 1$ where $m \neq 0$. If the lines $y = 0$ and $y = mx + 1$ form an angle of 30°, what angles are formed from the line $y = 0$ and $y = \frac{-1}{m}x - 1$?

Problem 1.10 Suppose angles $\angle A$ and $\angle B$ are complementary, with ratio of angles $\angle A : \angle B = 7 : 11$. What is the ratio of the angles supplementary to $\angle A$ and supplementary to $\angle B$?

1.2 Quick Response Questions

Problem 1.11 Let m and n be a pair of parallel lines and let transversal l cut across the parallel lines as shown in the figure below.

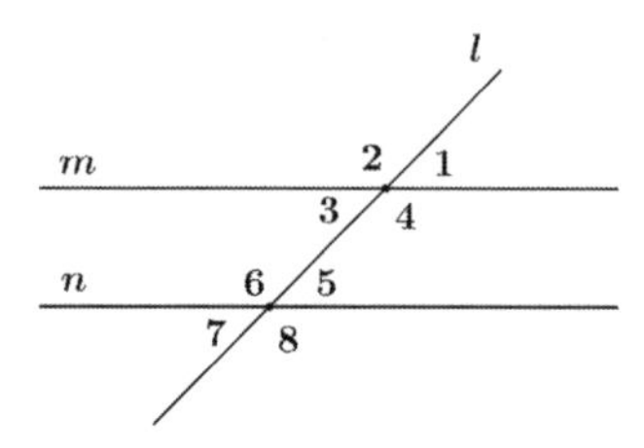

Which of the following angles are corresponding angles?

(A) $\angle 1 = \angle 3$
(B) $\angle 3 = \angle 7$
(C) $\angle 2 = \angle 8$
(D) $\angle 5 = \angle 7$

Problem 1.12 Let m and n be a pair of parallel lines and let transversal l cut across the parallel lines as shown in the figure below.

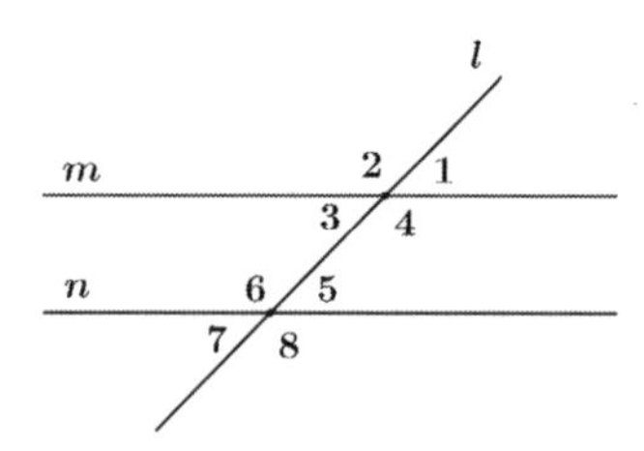

If $\angle 1 = 42^\circ$, what is the measure of $\angle 7$?

Problem 1.13 Let m and n be a pair of parallel lines and let transversal l cut across the parallel lines as shown in the figure below.

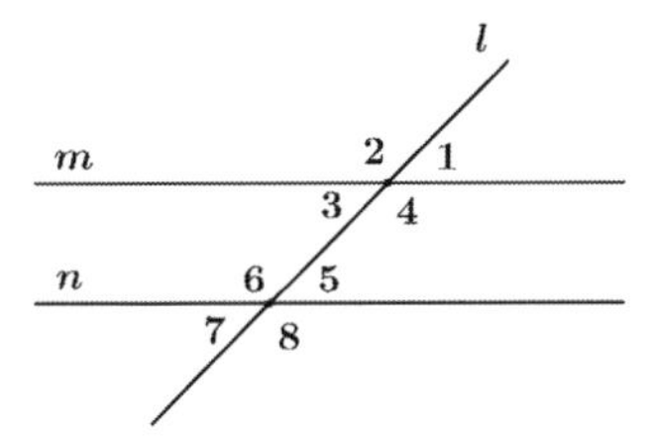

If $\angle 8 = 132°$, what is the measure of $\angle 3$?

Problem 1.14 In $\triangle ABC$, $\angle B = 28°$, and $\angle C = 37°$. What is the degree measure of $\angle A$?

Problem 1.15 A line goes through points $(-2,7)$ and $(-10,5)$. What is the slope of the line? Give your answer as a decimal.

Problem 1.16 Let $A = (x_A, 4)$ and $B = (9,7)$. What is x_A if $\overline{AB}$ is perpendicular to the line $y = -\frac{2}{3}x + 8$?

Problem 1.17 Consider the points $A = (8,7)$ and $B = (-12,9)$. What is the x coordinate of the midpoint of $\overline{AB}$?

Problem 1.18 Which of the following lines is not parallel to the line given by the equation $y = \frac{3}{5}x + 12$?

(A) $3x - 5y = 60$
(B) $5x - 3y = 36$
(C) $10y - 6x = -30$
(D) All three lines are parallel

Problem 1.19 Which of the following lines is perpendicular to $y = 8x - 4$?

(A) $8y = x + 4$
(B) $8y - x = 14$
(C) $8y + x = 12$
(D) None of the above

Problem 1.20 Which of the following is the equation of a line with slope 7 and y-intercept 31?

(A) $y = 31x + 7$
(B) $y = -7x - 31$
(C) $y = 7x + 31$
(D) $y = -31x + 7$

1.3 Practice Questions

Problem 1.21 Find the equation of the line that is perpendicular to $y = 2x - 3$ when $y = 3$.

Problem 1.22 $M = (-1, 2)$ is the midpoint of line segment $\overline{AB}$ with $A = (4, 3)$. What is B?

Problem 1.23 Points A, B, C are all on the same line. $A = (1, 3)$, $B = (3, k)$, and $C = (k, 7)$ for some number k. What are the possible values of k?

Problem 1.24 Triangle ABC has vertices $A = (0, 0)$, $B = (4, 0)$, and $C = (x_C, \sqrt{3})$. If $\angle ACB = 90^\circ$, what are the possible values for x_C?

Problem 1.25 Let O denote the origin $(0, 0)$ and points A, B, C are chosen such that $\angle BOC = 2 \times \angle AOB$ and $\angle COA = \angle BOC - 15^\circ$. What is $\angle AOB$?

Problem 1.26 Consider the diagram below, where l and m are parallel but the drawing is not necessarily to scale.

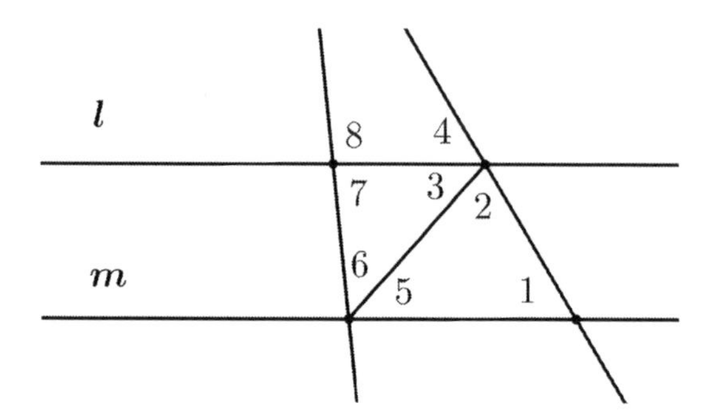

Suppose that $\angle 1 = 60°, \angle 5 = 50°, \angle 8 = 105°$. In class it was found that $\angle 6 = 55°$. Find the measure of the remaining angles.

Problem 1.27 Prove that sum of two interior angles in a triangle is equal to the exterior angle for the third side. That is, show that $\angle 1 + \angle 2 = \angle 3$ in the diagram below.

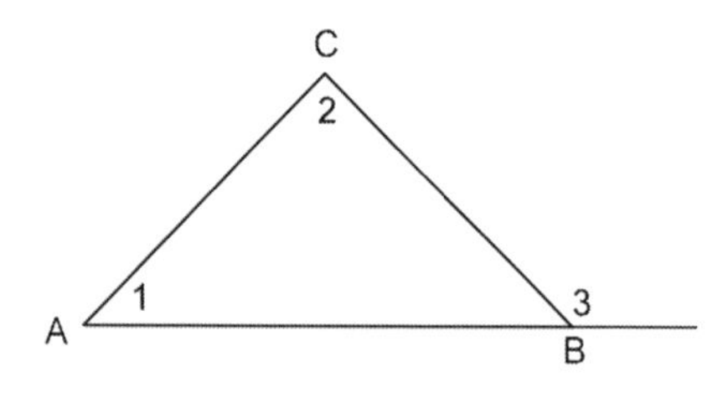

Problem 1.28 Suppose you have $\triangle ABC$ with angles $\angle A, \angle B, \angle C$. If $\angle A$ and $\angle C$ are complementary and $\angle B$ is three times $\angle A$, what is $\angle C$?

Problem 1.29 The line $y = mx$ (with $m > 0$) forms a 20° angles with the x-axis. The line $y = x - 2$ forms a 45° angle with the y-axis. The lines $y = mx$ and $y = x - 2$ meet and form an acute angle of K°. What is K?

Problem 1.30 Suppose angles $\angle A$ and $\angle B$ are supplementary, with ratio of angles $\angle A : \angle B = 7 : 11$. What is the ratio of the angles supplementary to $\angle A$ and supplementary to $\angle B$?

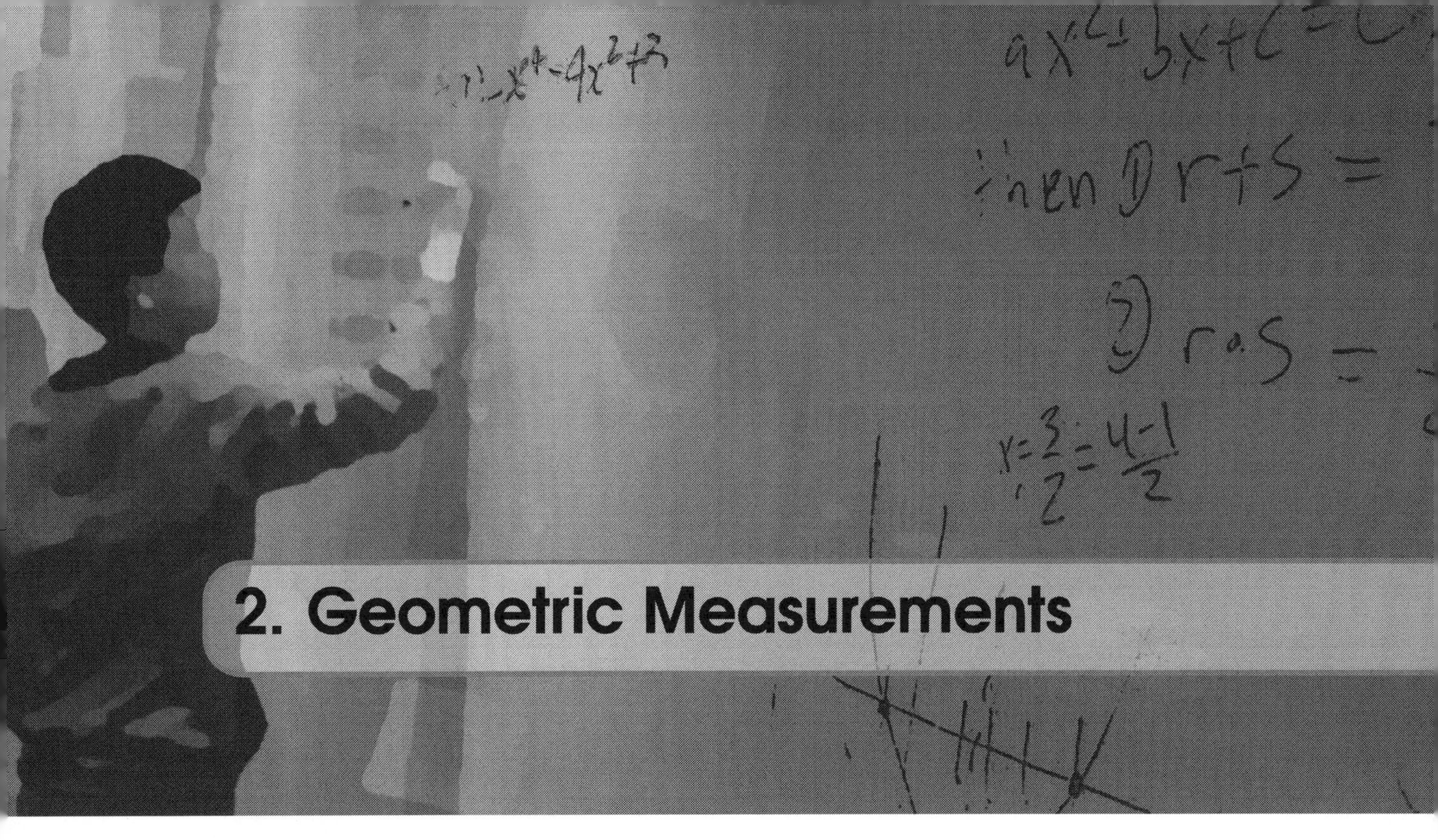

2. Geometric Measurements

Basic Measurements

- Perimeter of a square of side s: $P = 4 \times s$.
- Area of a square of side s: $A = s^2$.
- Perimeter of a rectangle of length l and width w: $P = 2 \times (l + w)$.
- Area of a rectangle of length l and width w: $A = l \times w$.
    - We can understand this formula by breaking a rectangle into small squares.

Pythagorean Theorem

- In a right triangle $\triangle ABC$ with $\angle C = 90^\circ$ and $AB = c, AC = b, BC = a$ we have $c^2 = a^2 + b^2$.
- Conversely, in a triangle $\triangle ABC$ with $AB = c, AC = b, BC = a$, if $c^2 = a^2 + b^2$ then $\triangle ABC$ is a right triangle.

Distance Formula

- The length of line segment $\overline{AB}$ where $A = (x_A, y_A)$ and $B = (x_B, y_B)$ is given by

$$\sqrt{(x_A - x_B)^2 + (y_A - y_B)^2}.$$

- Note the similarity to the Pythagorean theorem.

2.1 Example Questions

Problem 2.1 A rectangle is divided into 4 smaller rectangles by two lines, as shown. The perimeters of three of these rectangles are 12, 14, and 14. Find the perimeter of the remaining (shaded) rectangle.

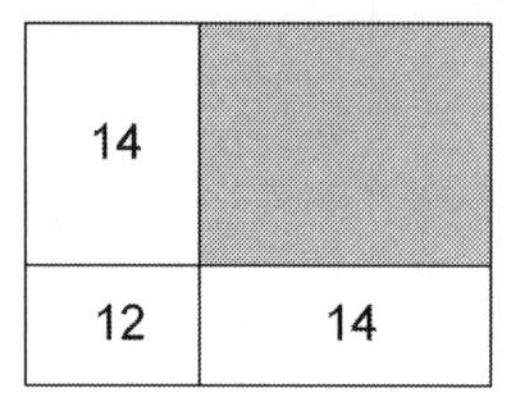

Problem 2.2 A rectangle is divided into 5 squares, as shown in the diagram.

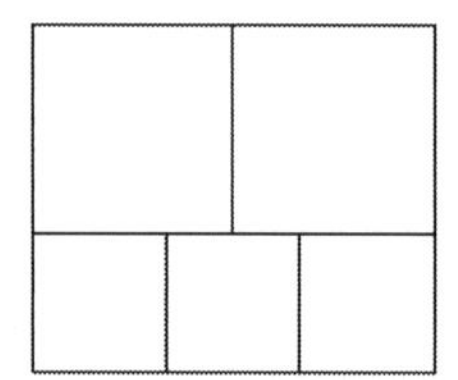

Given that the area of one bigger square is 16 in^2 more than that of one smaller square, find the area of the whole rectangle.

Problem 2.3 Arrange several equilateral triangles and rhombi, all of whose side lengths are 2 cm, to form a long parallelogram, as shown in the diagram.

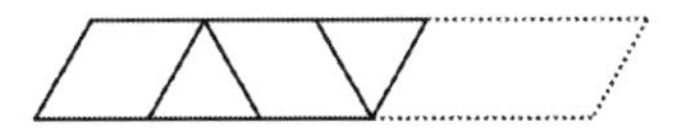

Assume the perimeter of the long parallelogram is 368 cm, how many equilateral triangle and rhombi are there?

Problem 2.4 Prove the Pythagorean Theorem using the diagram below:

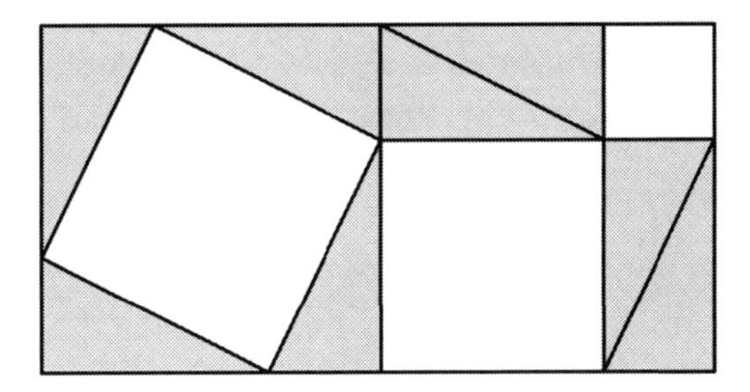

Problem 2.5 The lines $-4x+3y=2$, $-4x+3y=27$, $3x+4y=-14$, and $3x+4y=11$ intersect at the points $(-6,1)$, $(-3,5)$, $(-2,-2)$, and $(1,2)$ to form a quadrilateral.

Show that this quadrilateral is a square.

Problem 2.6 Review of areas of parallelograms, triangles, and trapezoid. Try to understand all of these using the area of a rectangle as a starting point.

(a) The area of a parallelogram is bh.

(b) The area of a triangle is $\frac{1}{2}bh$.

(c) The area of a trapezoid is $\frac{b_1+b_2}{2}h$.

Problem 2.7 Prove $(a+b)^2 = a^2 + 2ab + b^2$ geometrically.

Problem 2.8 Find the point on the line $y = 2x + 5$ that is the closest to the origin.

Problem 2.9 A big rectangle is divided into 6 squares of different sizes, as shown.

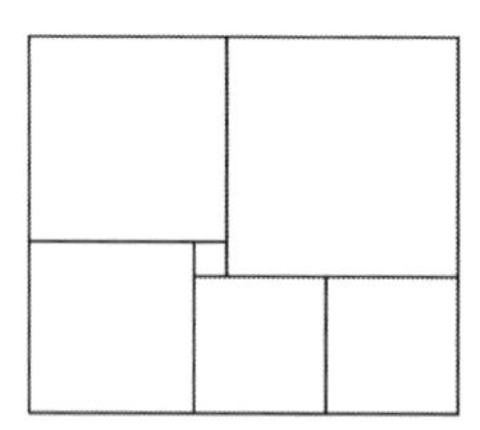

Given that the smallest square in the middle has area 1 cm^2, find the area of the big rectangle in square centimeters.

Problem 2.10 Suppose a triangle has vertices $(3,4),(4,7),(7,6)$. Find the area of the triangle.

2.2 Quick Response Questions

Problem 2.11 A rectangle is divided into 3 squares, as shown in the diagram. Given that the area of the rectangle is 294, find the length of the side of one of the small squares.r

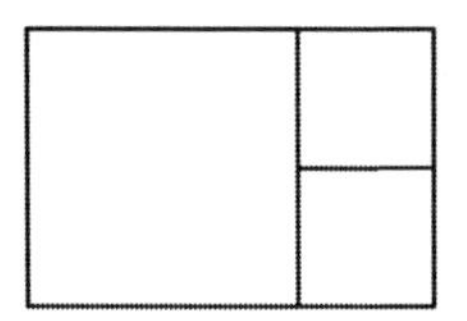

Problem 2.12 A trapezoid has area 75. If the height of the trapezoid is 5 and of of its bases has length 17, what is the length of the other base?

Problem 2.13 Which of the following triples could not be the lengths of the sides of a right triangle?

(A) $3,4,5$
(B) $5,12,13$
(C) $12,15,17$
(D) $7,24,25$

Problem 2.14 Consider the right triangle ABC, with $\angle A = 90^\circ$. If $AB = 21$ and $BC = 29$, what is AC?

Problem 2.15 Consider the points $A = (3,5)$, $B = (-4,5)$, $C = (-4,-2)$ and $D = (x_D, y_D)$ on the coordinate plane. If $ABCD$ is a rectangle, what is $x_D + y_D$?

Problem 2.16 What is the shortest distance from the point $(5,-3)$ to the line $x = 2$?

Problem 2.17 What is the distance between the points $(3,4)$ and $(8,16)$? Round your answer to the nearest tenth if necesary.

Problem 2.18 What is the distance between the points $(-2,3)$ and $(-10,-12)$? Round your answer to the nearest tenth if necessary.

Problem 2.19 Consider the points $A = (2,7)$, $B = (-2,10)$ and $C = (1,14)$. Is it true that $\triangle ABC$ is isosceles? (Recall isosceles means two sides of the triangle are equal.)

Problem 2.20 In the diagram, each side is perpendicular to its adhacent sides, and all small sides have equal length. Given that the perimeter of this diagram is 60, what is the area of the shape?

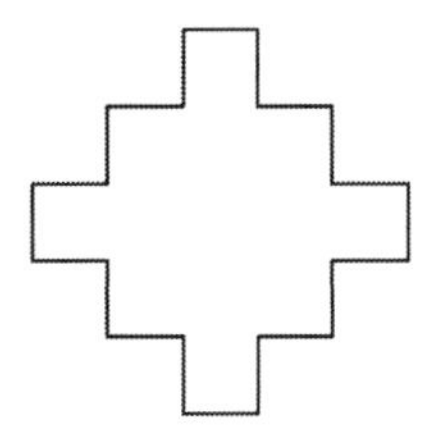

2.3 Practice Questions

Problem 2.21 A rectangle is divided into 4 smaller rectangles by two lines, as shown.

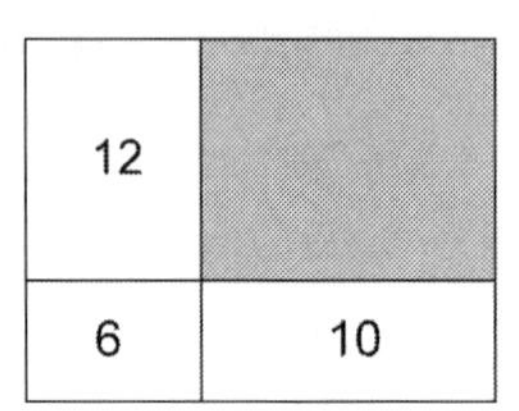

The areas of three of these rectangles are 6, 12, and 10. Find the area of the remaining (shaded) rectangle.

Problem 2.22 A rectangle is divided into 5 squares, as shown in the diagram.

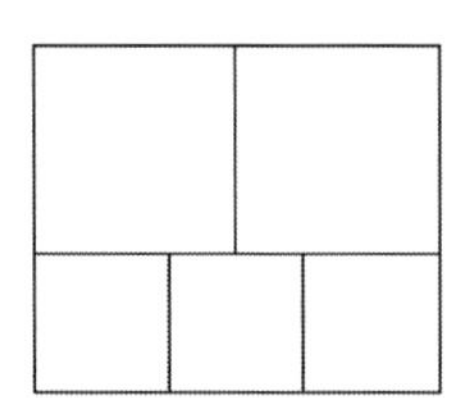

If the area of the full rectangle is 150 square inches, find the length and width of the rectangle.

Problem 2.23 Arrange several rhombi, all of whose side lengths are 3 cm, to form a long parallelogram, as shown in the diagram. Assume the perimeter of the long parallelogram is 108 cm. How many rhombi are there?

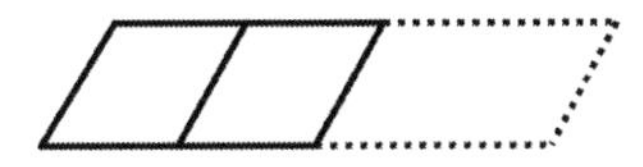

Problem 2.24 We call a triple (a,b,c) a Pythagorean triple if $a^2+b^2=c^2$. Prove algebraically that if (a,b,c) is a Pythagorean triple then so is $(k\times a, k\times b, k\times c)$ for $k>0$.

Problem 2.25 The lines $2x-y=-5$, $2x-y=5$, $x+2y=5$, and $x+2y=10$ intersect at 4 points to form a rectangle. What is the area of this rectangle?

Problem 2.26 Expanding on the areas we covered in class, try to understand the following formulas by using more basic shapes.

(a) The area of a trapezoid is also mh where m is the median of the trapezoid, which connects the midpoints of the two non-parallel sides.

(b) A quadrilateral whose diagonals (formed by connecting opposite vertices) are perpendicular is called a kite.

Prove that the area of a kite is $\frac{d_1 \cdot d_2}{2}$, where d_1, d_2 are the lengths of the diagonals.

Problem 2.27 Two rectangles and a square are assembled to form a big square as shown.

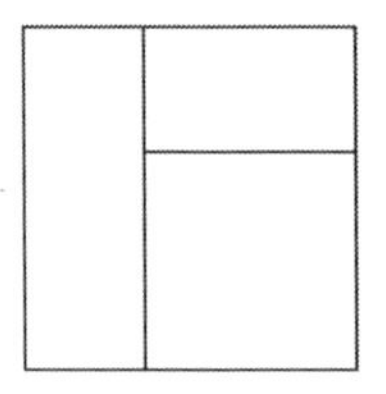

The area of the rectangles are 44 and 28. What is the area of the entire square?

Problem 2.28 Find the shortest distance from the point $(0,8)$ to the line $y = x$.

Problem 2.29 A big rectangle is divided into 6 squares of different sizes, as shown.

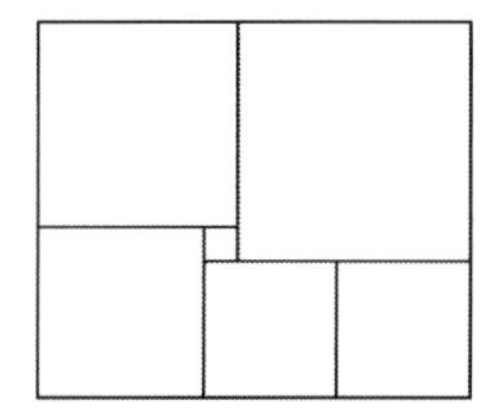

Given that the largest square in the upper right has area 196 cm^2, find the area of the big rectangle.

Problem 2.30 Let $A = (-2,0)$, $B = (0,2)$, $C = (2,2)$, $D = (-2,-2)$ and form quadrilateral $ABCD$. What is the area of $ABCD$?

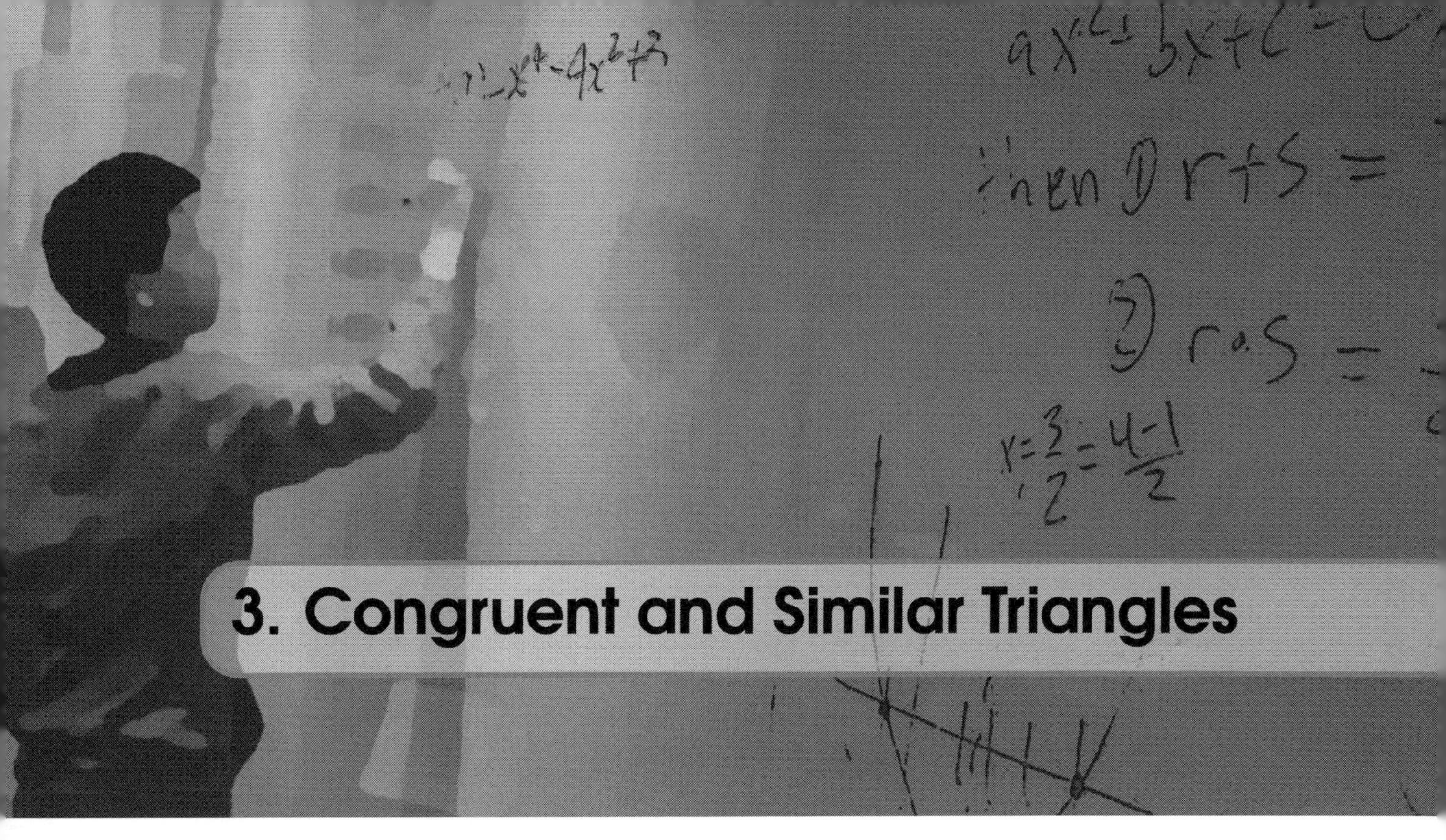

3. Congruent and Similar Triangles

Types of Triangles

- An *equilateral* triangle is a triangle made up of three equal sides and three equal angles.
- An *isosceles* triangle has two equal sides and the angles opposite those sides are equal.
- A *scalene* triangle has all different sides and all different angles.
- Note: In all of these examples, knowing about the sides OR the angles is enough. For example, if a triangle has two angles that are equal to each other, then the sides opposite those angles must be equal.

Congruent and Similar Triangles

- Two triangles are *congruent* if they are exactly the "same". More formally, this means all their sides and angles are equal.
- Two triangles are *similar* if they are the "same" except for possibly their size. More formally, this means their angles are equal and all their sides are in a common ratio (for example all the angles are the same but one triangle has sides that are all twice as long as the other triangle).
- In the diagram below $\triangle ABC$ is congruent to $\triangle DEF$ (written $\triangle ABC \cong \triangle DEF$). Triangle $\triangle GHI$ is NOT congruent to either $\triangle ABC$ or $\triangle DEF$. However, all three triangles are similar (written $\triangle ABC \sim \triangle DEF \sim \triangle GHI$).

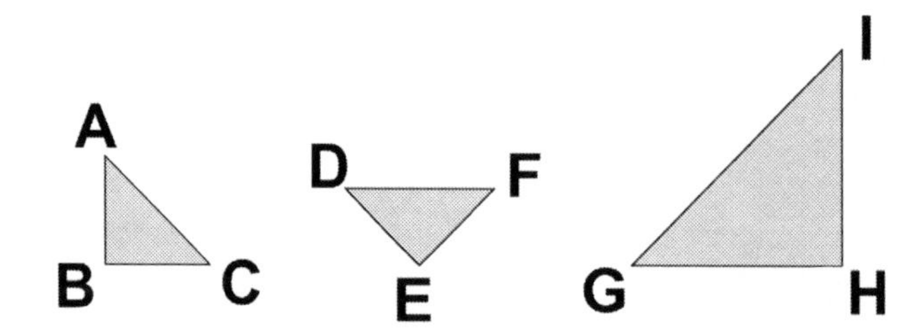

- If two triangles have all three sides the same length, then they are actually congruent. Think of making a triangle from three sticks, there is only one triangle you can make! This rule is often referred to as the **SSS** rule for congruence.
- Similarly, if two triangles are different sizes, but all the sides share a common ratio to each other they are similar. This rule is referred to as the **SSS** rule for similarity.

Lines in Triangles

- **Median:** The line from a vertex to the midpoint on the opposite side.
- **Altitude:** The line from a vertex which is perpendicular to the opposite side. Note this is sometimes referred to as the height of the triangle.
- **Angle Bisector:** The line from a vertex which bisects the angle at the vertex.

Two Special Right Triangles

- In $\triangle ABC$, if $\angle C = 90^\circ$, $\angle A = \angle B = 45^\circ$, then $AB = \sqrt{2}AC = \sqrt{2}BC$.
- In $\triangle ABC$, if $\angle C = 90^\circ$, $\angle A = 60^\circ$, and $\angle B = 30^\circ$, then $AB = 2AC$, $BC = \sqrt{3}AC$.

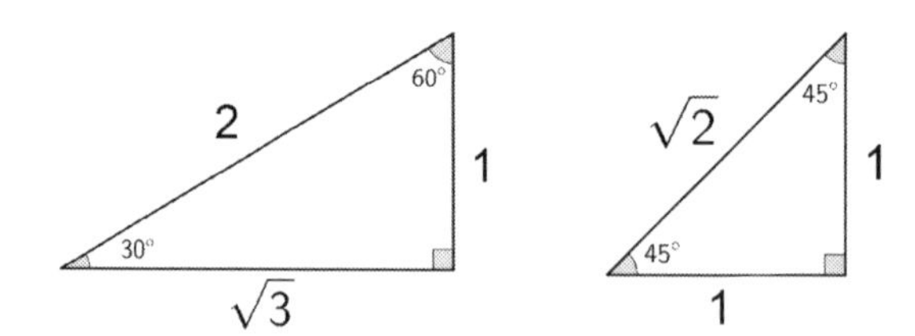

3.1 Example Questions

Problem 3.1 For each of the following "rules", state whether they work for proving congruence, similarity, both, or neither. If the rule does not work, give a counterexample.

(a) SAS (two sides and the angle between them)

(b) AAA (all three angles)

(c) ASA (two angles and the side between them)

(d) AAS (two angles and a side not between them)

(e) SSA (two sides and an angle not between them)

Problem 3.2 Let triangle ABC be equilateral triangle with side length 20. Let D be on side $\overline{AB}$ and E be on side $\overline{AC}$ such that $\overline{DE} \| \overline{BC}$. Assume triangle ADE and trapezoid $DECB$ have the same perimeter. What is the length of $\overline{BD}$?

Problem 3.3 Prove the converse of the Pythagorean Theorem. Note, we have already proven the Pythagorean theorem, so we can use it in this proof!

Problem 3.4 Let $\triangle ABC$ with $AB = 12$, $BC = 16$, $AC = 20$. Let D, E, and F be the midpoints of, respectively, AB, BC, and AC. What is the area of $\triangle DEF$?

Problem 3.5 A 'Golden Triangle' is an isosceles triangle with angles in ratio $1:2:2$.

(a) Find the angles in a golden triangle.

(b) Find the ratio of the sides in a golden triangle. The diagram shown below with golden triangle ABC may help.

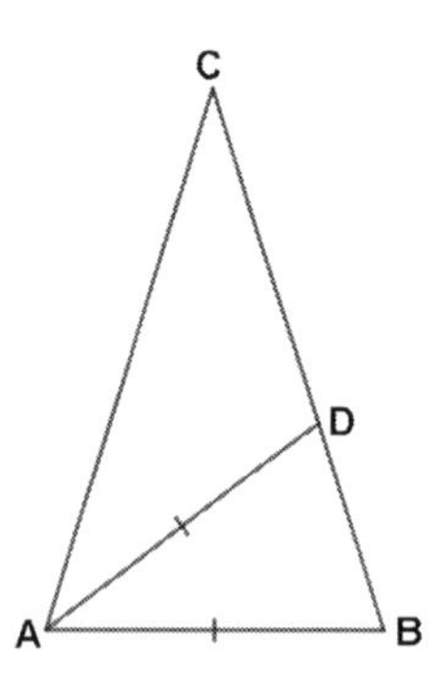

Problem 3.6 What is the ratio of the sides of a triangle with angles $30°$, $60°$, and $90°$?

Problem 3.7 Given that one of the angles of the triangle with sides $(5,7,8)$ is $60°$, show that one of the angles of the triangle with sides $(3,5,7)$ is $120°$.

Problem 3.8 Suppose A and B are two points. Describe the set of points that are equal distance from A as from B.

Problem 3.9 $\triangle ABC$ is an equilateral triangle with with $A = (0,2)$ and $B = (\sqrt{3},1)$. Find all possible coordinates for C.

Problem 3.10 What is the side length of the largest equilateral triangle $\triangle AEF$ that can fit inside square $ABCD$ with side length 1? For reference, $\triangle AEF$ is shown in the diagram, with $BE = DF$.

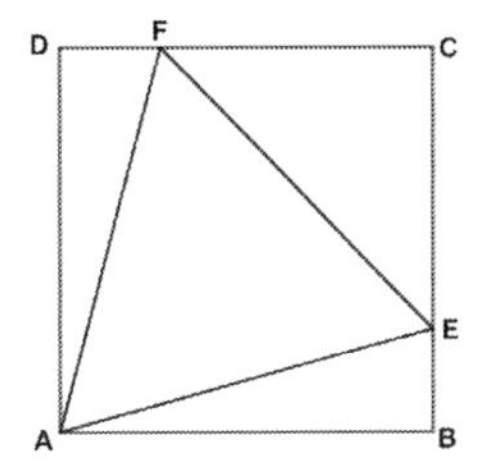

3.2 Quick Response Questions

Problem 3.11 Let ABC be a triangle with $AB = 15$, $\angle A = 32^\circ$, and $\angle B = 49^\circ$. Let DEF be a triangle with $DE = 45$. $\angle D = 32^\circ$ and $\angle F = 99^\circ$. Is it true that $\triangle ABC \sim \triangle DEF$?

Problem 3.12 Let ABC be a triangle with $AB = 12$, $BC = 15$, and $\angle A = 28^\circ$. Let DEF be a triangle with $DE = 24$, $DF = 30$, and $\angle E = 28^\circ$. Does it have to be true that $\triangle ABC \sim \triangle DEF$?

Problem 3.13 The sides of an equilateral triangle are 4 cm long. An altitude of this triangle is $A\sqrt{B}$, what is $A+B$?

Problem 3.14 Consider points $A = (1,3)$ and $B = (4,2)$. Is it true that $C = (3,5)$ is a point on the perpendicular bisector of AB?

Problem 3.15 $\triangle ABC \sim \triangle DEF$. If $AB = 20$, $BC = 12$, $DE = x+2$, and $EF = x-2$, what is x?

Problem 3.16 In $\triangle ABC$, $\angle A = \angle B$, $AB = x+1$, $BC = 3x-4$ and $AC = 2x+1$. What is AB?

Problem 3.17 Let $ABCD$ be a parallelogram, and let E be the interestction of its diagonals. Which of the following is not true?

(A) $\triangle AED \cong \triangle CEB$
(B) $\triangle AEB \cong \triangle CED$
(C) $\triangle ADB \cong \triangle ABC$
(D) $\triangle ABC \cong \triangle CDA$

Problem 3.18 Let ABE be a triangle and C, D, be points on BE, such that $AC = AD$ and $BD = CE$. Is it true that $AB = AE$?

Problem 3.19 Consider quadrilateral $ABCD$, with $AB || CD$ and $AB = CD$. Which can be used to show that $\triangle ABC \cong \triangle CDA$?

(A) SSS
(B) SAS
(C) ASA
(D) AAS

Problem 3.20 If $\triangle ABC \sim \triangle DEF$, with $AB = 32$, $BC = 12$, $CA = 16$ and $DE = 24$. What is the perimeter of $\triangle DEF$?

3.3 Practice Questions

Problem 3.21 Another common rule used for proving triangles are congruent and similar is called Hypotenuse-Leg (H-L). State the details of this rule and explain why it is true.

Problem 3.22 Let triangle ABC be equilateral triangle. Let D be on side $\overline{AB}$ and E be on side $\overline{AC}$ such that $\overline{DE} \| \overline{BC}$. Let the perimeter of triangle ADE to be 12 and the perimeter of trapezoid $DECB$ to be 16. Find the perimeter of triangle ABC.

Problem 3.23 Prove that two angles are equal in a triangle if and only if the opposite sides are equal. (Recall that to prove an if and only if you need to prove both directions!)

Problem 3.24 Let ABC be a triangle with area 100. Let D, E, and F be the midpoints of AB, BC, and AC, respectively. What is the area of $\triangle DEF$?

Problem 3.25 In a right triangle $\triangle ABC$ suppose $\angle B = 90°$ and $\angle A = 45°$. Suppose point D is on $\overline{BC}$ with $\angle ADB = 60°$ and $DC = 5$. Find the length of AB.

Problem 3.26 What is the ratio of sides of a triangle with angles $45°$, $45°$, and $90°$? Justify your answer.

Problem 3.27 In the following diagram $\angle ADB = 90^\circ$, $AD = 3$, $BD = 3$ and $BC = 6$. What is the measure of $\angle ABC$?

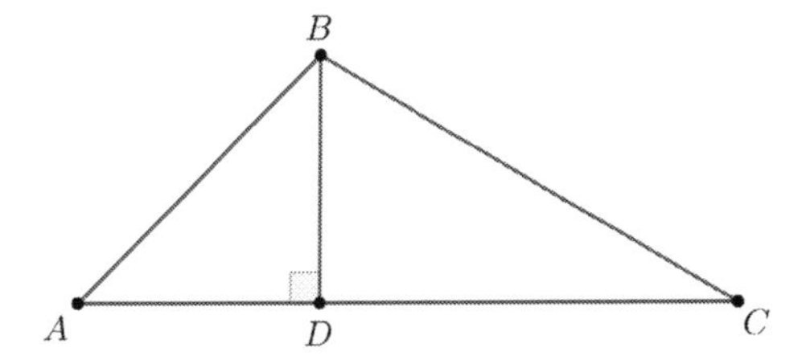

Problem 3.28 Consider points $A = (1,3)$ and $B = (4,2)$. If $\triangle ABC$ is isosceles with $AC = BC$, then C must lie on what line?

Problem 3.29 Consider points $A = (1,3)$ and $B = (4,2)$. If $\triangle ABC$ is isosceles triangle with $AB = AC$ and area 5, what are all possible values of C?

Problem 3.30 Draw the largest possible square inside an equilateral triangle, with one side of the square aligned with one side of the triangle. If the equilateral triangle has side length 6, find the side length of the square.

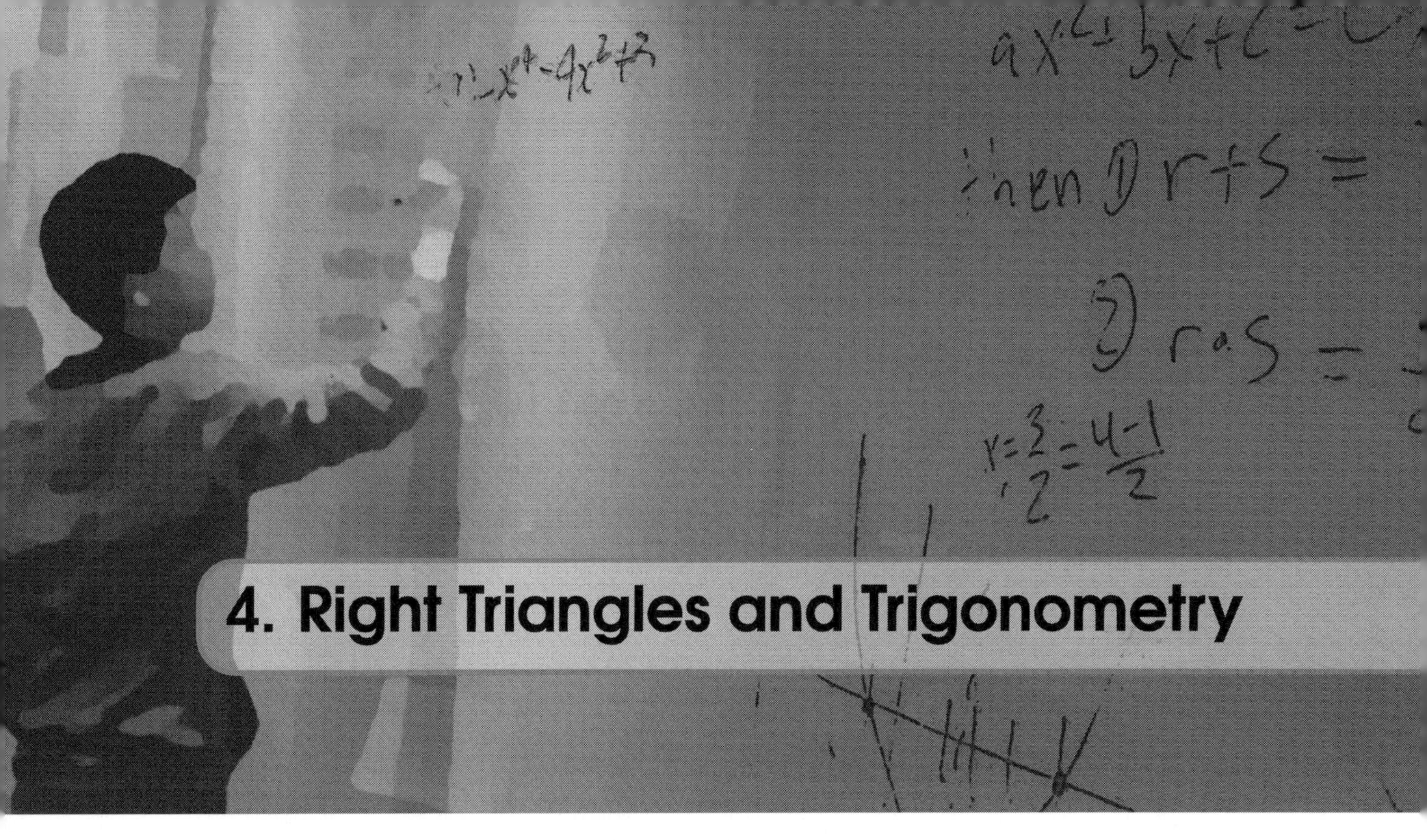

4. Right Triangles and Trigonometry

Two Special Right Triangles

- In $\triangle ABC$, if $\angle C = 90^\circ$, $\angle A = \angle B = 45^\circ$, then $AB = \sqrt{2}AC = \sqrt{2}BC$.
- In $\triangle ABC$, if $\angle C = 90^\circ$, $\angle A = 60^\circ$, and $\angle B = 30^\circ$, then $AB = 2AC$, $BC = \sqrt{3}AC$.

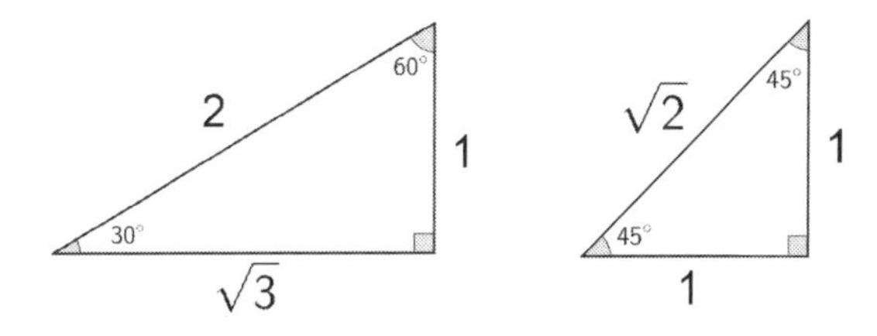

Sine, Cosine, and Tangent

- We'll denote angles today using the Greek letter θ (theta).
- In a right triangle, we will name the sides in reference to the angle θ, as in the diagram below.

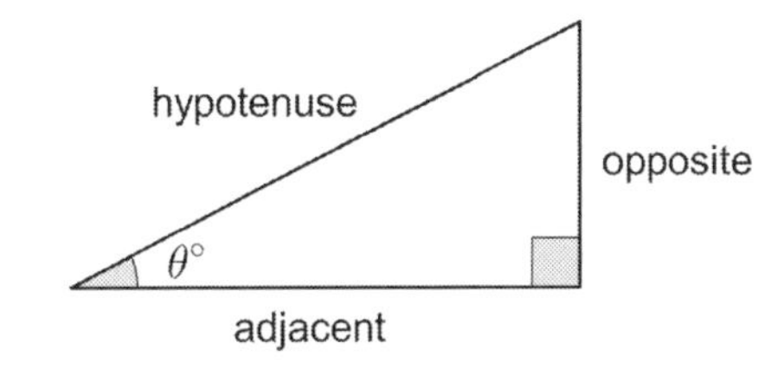

- Given a right triangle as above we let $\sin(\theta)$ ("sine theta") be the length of the opposite side divided by the length of the hypotenuse.
- Similarly, for a right triangle we let $\cos(\theta)$ ("cosine theta") be the length of the adjacent side divided by the length of the hypotenuse.
- Lastly, for a right triangle we let $\tan(\theta)$ ("tangent theta") be the length of the opposite side divided by the length of the hypotenuse.
- The mnemonic SOH-CAH-TOA is often helpful to remember the above relationships

$$\sin(\theta) = \frac{O}{H}, \quad \cos(\theta) = \frac{A}{H}, \quad \tan(\theta) = \frac{O}{A}$$

4.1 Example Questions

Problem 4.1 Similar triangles review. We use $a = BC, b = AC, C = AB$ to denote the sides of $\triangle ABC$.

(a) Let ABC be a right triangle with $\angle A = 30°$, $\angle C = 90°$, and $c = 4$. What are a and b?

(b) Let ABC be a triangle with $\angle A = 15°$, $\angle C = 90°$, $a = \sqrt{3} - 1$, $b = \sqrt{3} + 1$, and $c = 2\sqrt{2}$. Let DEF be a triangle with $\angle D = 15°$, $\angle F = 90°$, and $DE = 12$. What are EF and DF?

Problem 4.2 The following triangles are not necessarily drawn to scale. For each of them calculate $\sin(\theta), \cos(\theta), \tan(\theta)$.

(a) 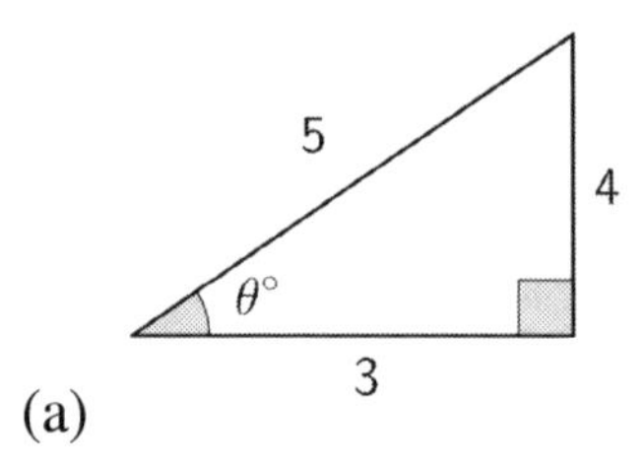

(b) 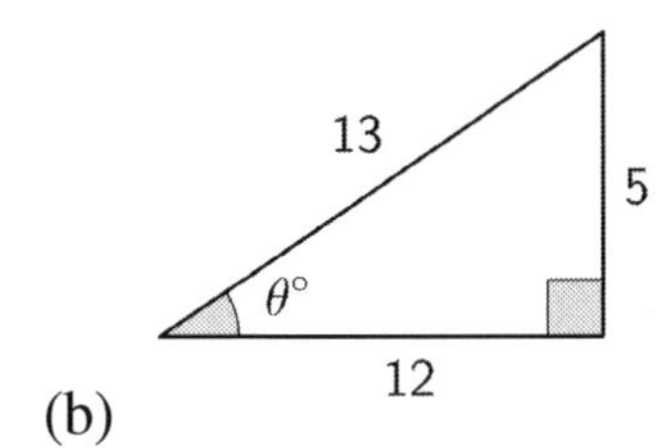

Problem 4.3 Calculate
(a) $\sin(60^\circ)$

(b) $\cos(60^\circ)$

(c) $\tan(60^\circ)$

Problem 4.4 Suppose you have an angle θ in a right triangle with $\cos(\theta) = \frac{4}{5}$.
(a) Calculate $\sin(\theta)$.

(b) Calculate $\tan(\theta)$.

Problem 4.5 Suppose $\triangle ABC$ is a right triangle with $\angle C = 90^\circ$. Let $BC = a$, $AC = b$, $AB = c$. Find a, b, c for each of the triangles below using the given information.
(a) $\sin(\angle A) = \frac{4}{5}$ and $c = 25$.

(b) $\tan(\angle A) = 1$ and $c = 4$.

Problem 4.6 Verify that $\tan(\theta) = \dfrac{\sin(\theta)}{\cos(\theta)}$ for $\theta = 30^\circ$.

Problem 4.7 Verify that $\sin^2(\theta) + \cos^2(\theta) = 1$ for $\theta = 60^\circ$. Note: $\sin^2(\theta) = (\sin(\theta))^2$.

Problem 4.8 Let ABC be a right triangle with $\angle C = 90^\circ$. If $\sin(\angle B) = \dfrac{7}{25}$, and $BC = 48$ what is the length of the height from vertex C?

Problem 4.9 Suppose you know you are 10 feet from the base of a building. You also know that the angle from the ground to the top of the building is 80°. How tall is the building? Hint: $\tan(80^\circ) \approx 5.67$.

Problem 4.10 Suppose John's cup is a cylinder with diameter 5 cm and height 12 cm. Suppose he wants to fill his cup halfway, but the water fountain he uses to fill it only shoots water to a height of 4.8 cm. Can he fill his cup halfway? Hint: Find the exact height needed to fill the cup halfway.

4.2 Quick Response Questions

Problem 4.11 In right triangle ABC, $\angle B = 90°$, $AB = 7.5$ and $AC = 19.5$. What is BC? Round your answer to the nearest tenth if necessary.

Problem 4.12 Right triangle ABC is such that $\angle C = 90°$, $AB = 18$ and $\sin(\angle A) = \frac{2}{9}$. What is BC? Round your answer to the nearest tenth if necessary.

Problem 4.13 Let ABC be a right triangle with $\angle C = 90°$ and $\tan(\angle B) = \frac{3}{4}$. Which of the following is not true?

(A) $\sin(\angle A) = \frac{4}{5}$
(B) $\cos(\angle B) = \frac{4}{5}$
(C) $\tan(\angle A) = \frac{3}{4}$
(D) None, they are all true

Problem 4.14 In $\triangle ABC$, $AC = 23$, $\angle C = 90°$, and $\tan(\angle A) = 1.3$. What is BC?

Problem 4.15 Is it possible to construct a right triangle ABC with $\angle C = 90°$ such that $\sin(\angle A) = \frac{11}{13}$?

Problem 4.16 Is it possible to construct a right triangle ABC with $\angle C = 90°$ such that $\sin(\angle B) = \frac{9}{7}$?

Problem 4.17 Let $ABCD$ be a right trapezoid with $AB \| CD$, $\angle C = 90^\circ$, $\angle D < 90^\circ$, $AB = 20$, $DA = 10$ and $\sin(\angle D) = \frac{4}{5}$. What is the area of $ABCD$?

Problem 4.18 θ is an acute angle in a right triangle and $\sin(\theta) = \frac{12}{19}$. Which of the following is equal to $\cos(\theta)$?

(A) $\frac{12}{217}$
(B) $\frac{\sqrt{217}}{19}$
(C) $\frac{12}{\sqrt{217}}$
(D) $\frac{217}{361}$

Problem 4.19 A ladder that is 12 ft long is resting on a wall and it makes an angle of 35° with respect to the ground. If $\cos(35^\circ) \approx .8192$ How many feet away from the wall is this ladder? Round your answer to the nearest tenth.

Problem 4.20 You want to measure the height of a tree in your backyard. The angle of elevation from a point on the ground that is 50 ft away from the base of the tree to the top of the tree is 42°. If $\tan(42^\circ) \approx .9004$, what is the height of the tree? Round your answer to the nearest foot.

4.3 Practice Questions

Problem 4.21 Calculate
(a) $\sin(30^\circ)$

(b) $\cos(30^\circ)$

(c) $\tan(30^\circ)$

Problem 4.22 The following triangles are not necessarily drawn to scale. For each of them calculate $\sin(\theta), \cos(\theta), \tan(\theta)$.

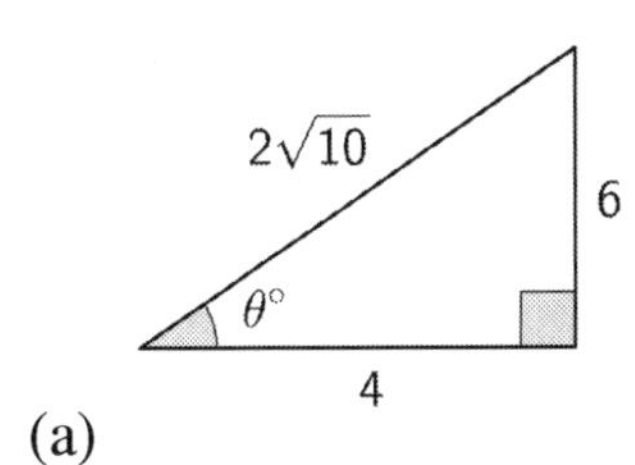

(a)

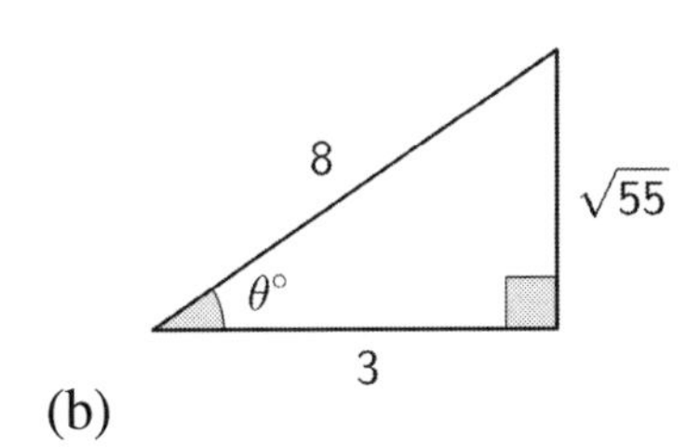

(b)

Problem 4.23 Suppose you have an angle θ in a right triangle with $\tan(\theta) = \frac{3}{2}$.

(a) Calculate $\sin(\theta)$.

(b) Calculate $\cos(\theta)$.

Problem 4.24 Suppose $\triangle ABC$ is a right triangle with $\angle C = 90^\circ$. Let $BC = a$, $AC = b$, $AB = c$. Find a, b, c for each of the triangles below using the given information.

(a) $\cos(\angle B) = \dfrac{5}{13}$ and $b = 24$.

(b) $\cos(\angle A) = \dfrac{1}{3}$ and $c = 6$.

Problem 4.25 Verify that $\tan(\theta) = \dfrac{\sin(\theta)}{\cos(\theta)}$ for $\theta = 60^\circ$.

Problem 4.26 Verify that $\sin^2(\theta) + \cos^2(\theta) = 1$ for $\theta = 45^\circ$.

Problem 4.27 Let ABC be a right triangle with $\angle C = 90^\circ$. If $\sin(\angle A) = \dfrac{8}{17}$, and $AB = 34$ what is the length of the height from vertex C?

Problem 4.28 Let ABC be a right triangle with $\angle C = 90^\circ$, and let D be the midpoint of BC. If $\tan(\angle ADC) = \dfrac{21}{10}$ and $AC = 42$, what is AB?

Problem 4.29 You are standing 5 ft away from a wall where a huge painting is being displayed. If you look down at an angle of 40° you are staring right at the bottom of the painting. If you look up at an angle of 65° you are staring right at the top of the painting. If $\tan(40^\circ) \approx .8391$, and $\tan(65^\circ) \approx 2.1445$, what is the height of the painting?

Problem 4.30 A pole is attached by two cables to the ground. Both cables are anchored from the same point on the ground. One cable makes a 75° with the ground and the other cable makes an angle of 55° with the ground. The distance between the points at which the cables are attached on the pole is 3 ft. Using the approximations $\tan(55^\circ) \approx 1.428$ and $\tan(75^\circ) \approx 3.732$, how tall is the pole?

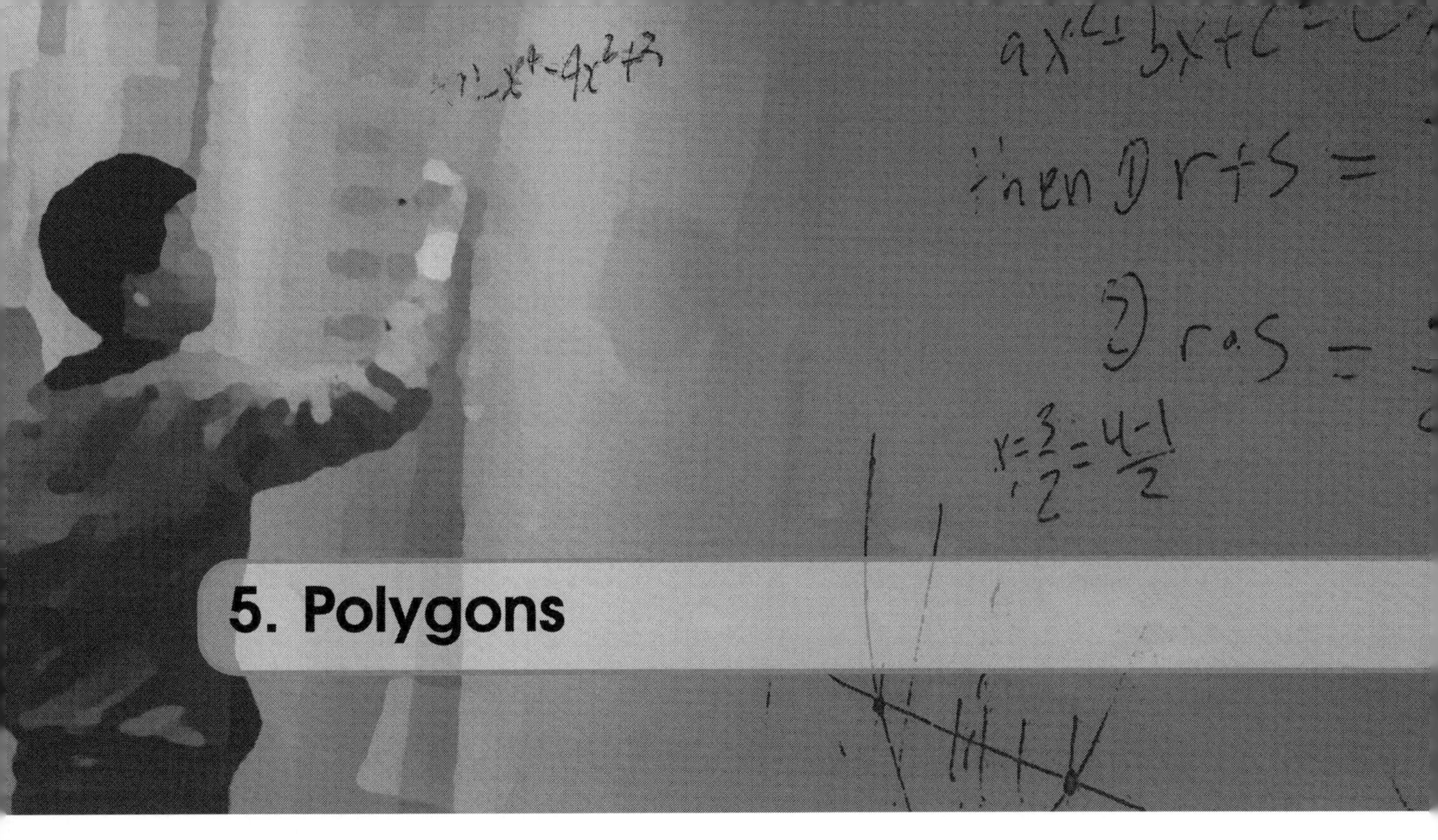

5. Polygons

Review and Definitions

- **Equilateral polygon:** A polygon that has all of its sides of the same length is called equilateral.
- **Equiangular polygon:** A polygon that has all of its internal angles the same size is called equiangular.
- **Regular polygon:** A polygon that is both equilateral and equiangular is called regular.
- We use $[ABC]$ to denote the area of triangle ABC, $[DEFG]$ to denote the area of quadrilateral $DEFG$, etc.

5.1 Example Questions

Problem 5.1 We've already shown that the interior angles of a triangle add up to 180°. Explain a general formula for the sum of the interior angles of a polygon with n sides.

Problem 5.2 Exterior Angles

(a) Prove that the exterior angles of a triangle add up to 360°.

(b) Explain a general formula for the sum of the exterior angles in a polygon with n sides.

Problem 5.3 Complete the following table about polygons with n sides: name, sum of interior angles, sum of exterior angles, and measure of each angle in case of regular polygon. All angles are in degrees. Justify your answers. Keep the chart for your own reference.

n	Name	Int. Angle Sum	Ext. Angle Sum	Each Angle (if regular)
3	Triangle			
4				
5				
6				
7	Heptagon			
8				
9	Nonagon			
10				
12	Dodecagon			
20	Icosagon			

Problem 5.4 Four non-overlapping regular plane polygons all have sides of length 1. The polygons meet at a point A in such a way that the sum of the four interior angles at A is 360°. Among the four polygons, two are squares and one is a triangle. What is the perimeter of the entire shape?

Problem 5.5 Consider the quadrilateral $ABCD$ shown below.

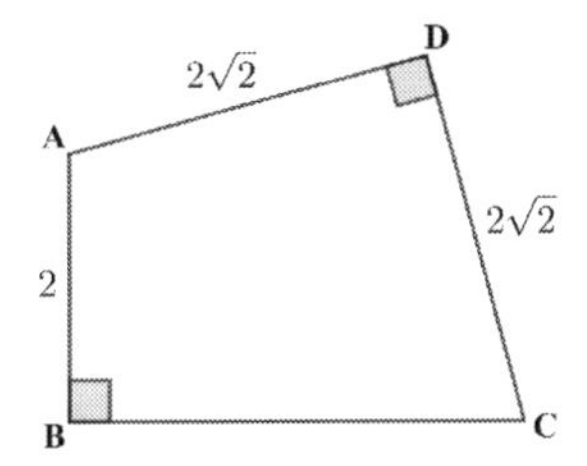

Find the missing side BC as well as the measures of $\angle A$ and $\angle C$.

Problem 5.6 Area of Triangles for SAS

(a) Find the area of $\triangle ABC$ if $BC = 5$, $AC = 6$, and $\angle C = 60^\circ$.

(b) Extend your method in part a) to give a formula for the area of $\triangle ABC$ if you know $BC = a$ and $AC = b$ and $\angle C = \theta$.

Problem 5.7 A square is formed by putting 4 congruent isosceles right triangles in the corners. The shaded square is the region not covered by the triangles as shown in the diagram below.

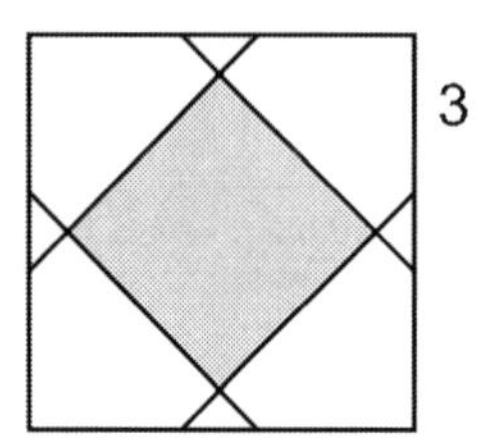

Caution: 3 is not the side length of the isosceles triangle. What is the area of the shaded square?

Problem 5.8 In parallelogram $ABCD$ as shown, $BC = 10$. Triangle BCE is a right triangle where $\overline{BE}$ is the hypotenuse, and $EC = 8$. Given that $[ABG] + [CDF] - [EFG] = 10$, find the length of $\overline{CF}$.

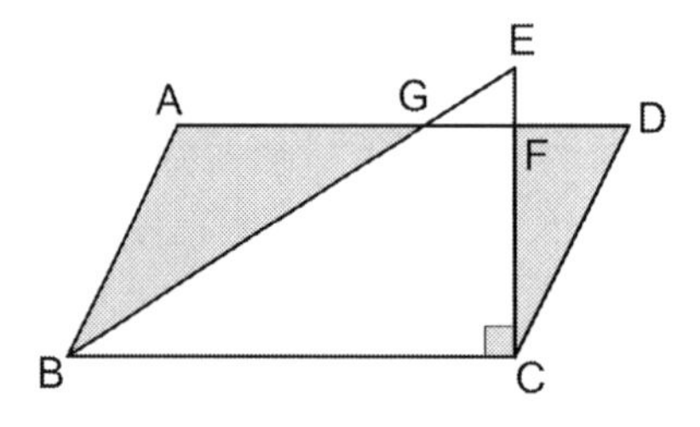

Problem 5.9 In equiangular octagon $ABCDEFGH$, $AB = CD = EF = GH = 6\sqrt{2}$ and $BC = DE = FG = HA$. Given the area of the octagon is 184, compute the length of side BC.

Problem 5.10 Inscribed Regular Polygons

(a) Find the area of the largest equilateral triangle that fits inside a circle of radius 1. Note, this triangle is said to be inscribed in a circle of radius 1.

(b) Extend your method from part (a) to give a general formula for the area of a regular polygon with n sides inscribed inside a circle of radius 1.

5.2 Quick Response Questions

Problem 5.11 Is the following an equiangular polygon that is not equilateral? (This is not a trick question.)

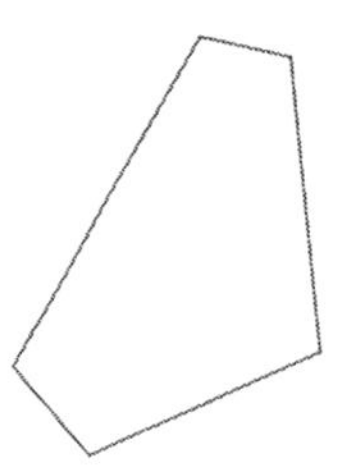

Problem 5.12 Is the following an equilateral polygon that is not equiangular? (This is not a trick question.)

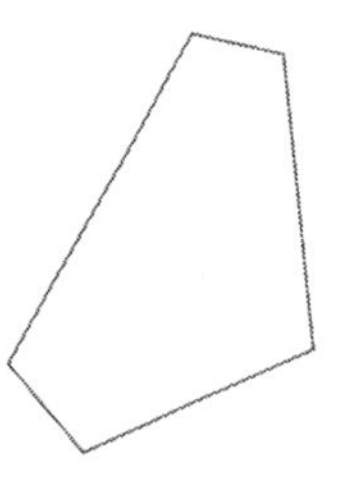

Problem 5.13 Four of the angles of a pentagon add up to 410°. What is the measure of the fifth angle?

Problem 5.14 A regular pentadecagon is a polygon with fifteen sides. What is the sum of its exterior angles?

Problem 5.15 A regular n-sided polygon has exterior angles of m degrees each. What is m?

(A) $m = 360$
(B) $m = 180(n-2)$
(C) $m = \frac{360}{n}$
(D) $m = 360(n-2)$

Problem 5.16 A regular heptagon of area 84 is shown in the diagram below.

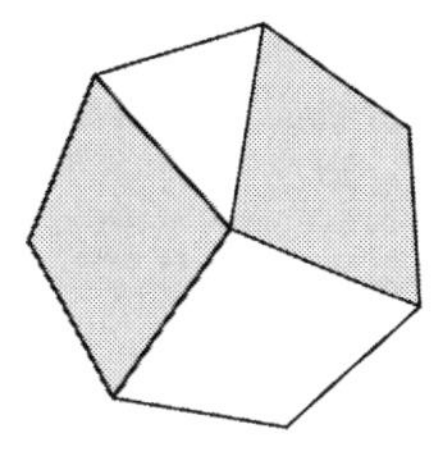

What is the shaded area?

Problem 5.17 Which of the following is NOT an example of a regular polygon which has integer interior angles?

(A) Decagon (10 sides)
(B) Dodecagon (12 sides)
(C) Tetradecagon (14 sides)
(D) Pentadecagon (15 sides)

Problem 5.18 If $BE = 10$, $AD = 8$ and the area of right triangle BCE is equal to the area of parallelogram $ABCD$ in the diagram below,

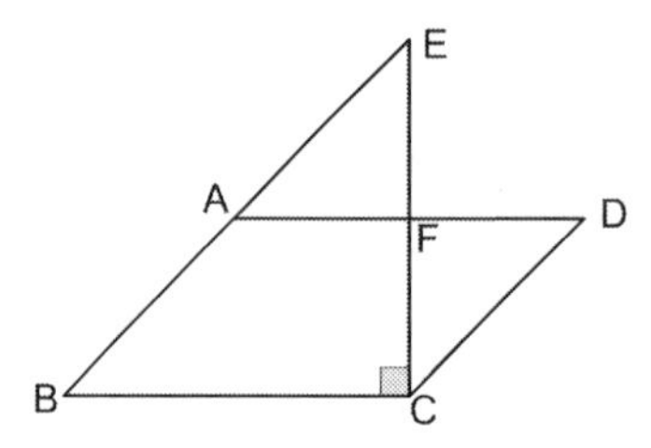

what is the area of trapezoid $ABCF$?

Problem 5.19 Find the area of the largest regular hexagon that fits in an equilateral triangle of area 108.

Problem 5.20 As shown in the diagram below, the equiangular convex hexagon $ABCDEF$ has $AB = 1$, $BC = 4$, $CD = 2$, and $DE = 4$. $[ABCDEF] = \frac{P\sqrt{Q}}{R}$. What is $P+Q+R$?

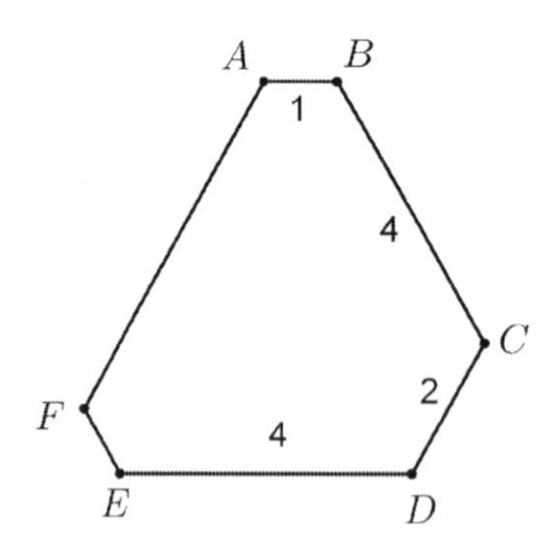

5.3 Practice Questions

Problem 5.21 Squares (regular quadrilaterals) can be repeatedly used to cover the entire plane with no gaps (called a tiling) as shown below.

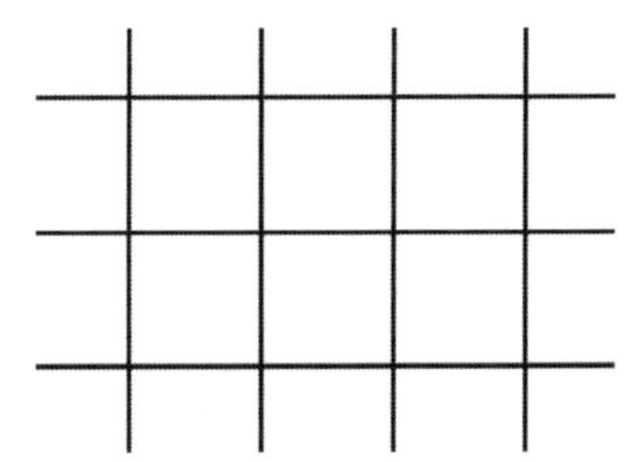

How many other regular polygons could be used instead of squares and also tile the plane? Which polygons are they?

Problem 5.22 Consider the exterior angles for a regular n-sided polygon. For how many regular polygons are these exterior angles an integer?

Problem 5.23 Let $ABCDEFGH$ be a regular octagon, and let $GHIJKL$ be a regular hexagon. Find all possible values of measure of $\angle IAH$. Note this means that $\overline{GH}$ is shared with both the octagon and the hexagon.

Problem 5.24 Three non-overlapping regular plane polygons all have sides of length 1. The polygons meet at a point A in such a way that the sum of the three interior angles at A is 360°. Thus the three polygons form a new polygon P (not necessarily convex) with A as an interior point. Suppose two of the polygons are pentagons. Find the perimeter of P.

Problem 5.25 Consider the quadrilateral $ABCD$ shown below.

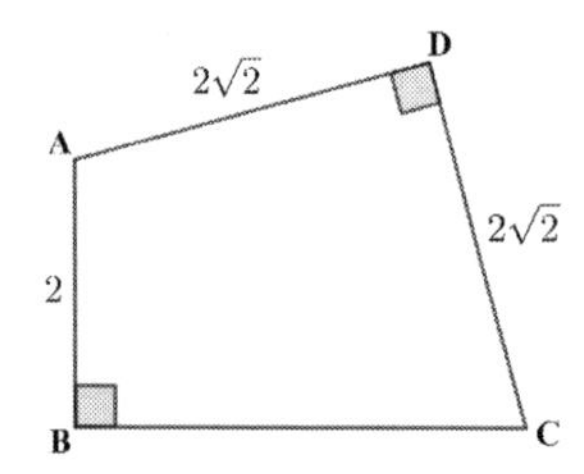

Find the area of the quadrilateral.

Problem 5.26 $\sin(75^\circ) = \dfrac{1+\sqrt{3}}{2\sqrt{2}} = \dfrac{\sqrt{2}+\sqrt{6}}{4}$. Using this information to calculate the area of $\triangle ABC$ if $BC = 8$, $AC = 5\sqrt{2}$, and $\angle C = 75^\circ$.

Problem 5.27 In the diagram, $\triangle ABC, \triangle DEF$ are two congruent isosceles right triangles. Given that $AB = 6, EC = 2$, find the area of the shaded region.

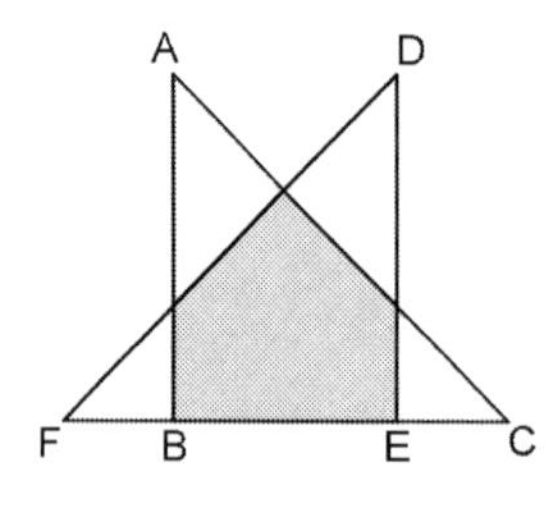

Problem 5.28 In the square shown in the diagram, the side length is 6, and the sum of the areas of the two shaded regions is 12. Find the area of quadrilateral $ABCD$.

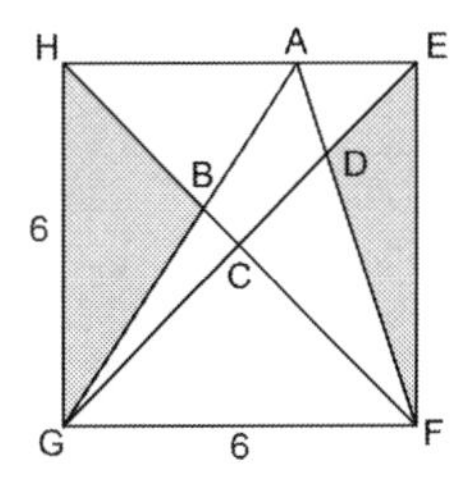

Problem 5.29 In equiangular octagon $ABCDEFGH$, $AB = CD = EF = GH$ and $BC = DE = FG = HA$. Argue that $\overline{AB} \perp \overline{CD}$.

Problem 5.30 Inscribed Areas

(a) Inscribe a regular hexagon inside a circle of radius 1. What is the area of the hexagon?

(b) Inscribe a regular dodecagon inside a circle of radius 1. What is the area of the dodecagon?

6. Circles

Basic Definitions

- A *circle* is a collection of points of equal distance (called the *radius*) from a set point (called the *center*).
- Given two points A, B on a circle, the segment $\overline{AB}$ is called a *chord*.
- If a chord AB contains the center of the circle, call AB a *diameter*.
- The portion of a circle that lies above or below a chord AB is called an *arc*. If the arc is more than half a circle it is called a *major arc*, less than half a circle is called a *minor arc*, and half a circle is called a *semicircle*. The arc will be denoted $\widehat{AB}$.
- Suppose $\widehat{AB}$ is an arc on a circle with center O. The *angular size* of the arc $\widehat{AB}$ is equal to the angle $\angle AOB$ (which is referred to as a *central angle*).
- Note: a full circle is 360°, a half circle is 180°, and a quarter circle is 90°.
- Given a central angle $\angle AOB$ from an arc $\widehat{AB}$, the figure contained between the arc $\widehat{AB}$ and the radii $\overline{OA}, \overline{OB}$ is called a *sector*.
- The diagram below has a few examples:

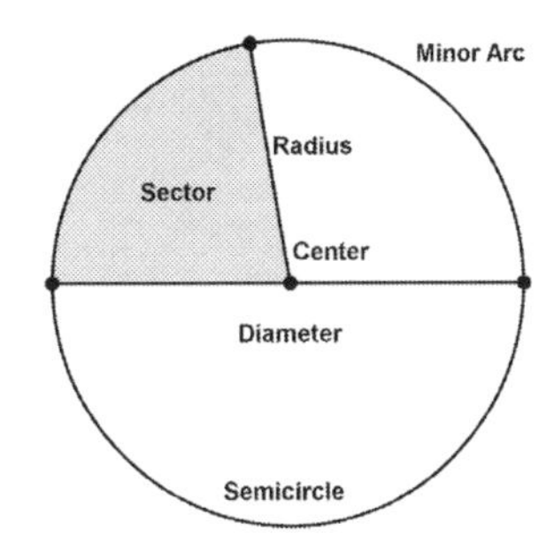

Measurements in Circles

- The area of a circle is given by πr^2 where r is the radius.
- The circumference of a circle is given by $2\pi r = \pi d$ where r, d are the radius and length of a diameter respectively.
- The area of a sector from arc $\widehat{AB}$ is given by $\dfrac{\theta}{360^\circ}\pi r^2$ where θ is the angular size of $\widehat{AB}$ (measured in degrees).
- In particular, the area of half a circle is $\frac{180^\circ}{360^\circ}\pi r^2 = \frac{1}{2}\pi r^2$ and the area of a quarter circle $\frac{90^\circ}{360^\circ}\pi r^2 = \frac{1}{4}\pi r^2$.
- Similarly, *arc length* of $\widehat{AB}$ (that is, the distance walking from A to B along the circle) is given by $\dfrac{\theta}{360^\circ}2\pi r$ where θ is the angular size of $\widehat{AB}$ (measured in degrees).
- In particular, the arc length of a semicircle is $\frac{180^\circ}{360^\circ}2\pi r = \frac{1}{2}2\pi r = \pi r$ and the arc length of a quarter circle is $\frac{90^\circ}{360^\circ}2\pi r = \frac{1}{4}2\pi r = \frac{1}{2}\pi r$.

Equations for Circles

- $x^2 + y^2 = r^2$ is a circle with radius r centered at the origin $(0,0)$.
- Translations:
  - Replacing x by $x - h$ in an equation shifts the graph h units right.
  - Replacing y by $y - k$ in an equation shifts the graph k units up.
- General Equation: $(x-h)^2 + (y-k)^2 = r^2$ is a circle with radius r centered at (h,k).

Theorems about Perpendiculars

- In a circle, a radius is perpendicular to a chord if and only if the radius bisects the chord.
- In a circle, the perpendicular bisector of a chord passes through the center of the chord.
- Similar to the above: Suppose a line going through a point P on the circle. The line is tangent to the circle if and only if the line is perpendicular to the radius of the circle.

Arcs and Angles

- If points A, B, P are on a circle, we call $\angle APB$ an *inscribed angle*. The measure of $\angle APB$ is half the angular size of arc $\widehat{AB}$ (where the arc does NOT contain P).

- Suppose two chords AC, BD intersect inside the circle at a point P as in the diagram below.

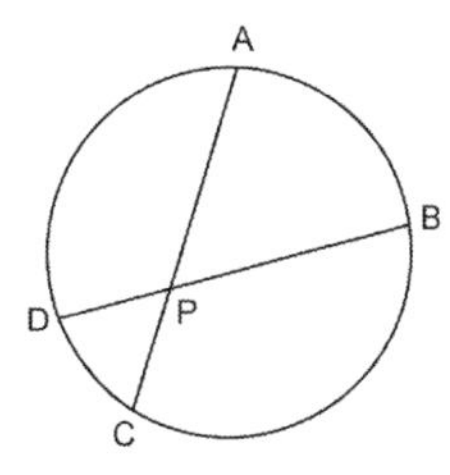

Then $\angle APB$ is half the sum of the angular sizes of arcs $\widehat{AB}$ and $\widehat{CD}$.

- Suppose the extension of two chords AC, BD intersect outside the circle at a point P as in the diagram below.

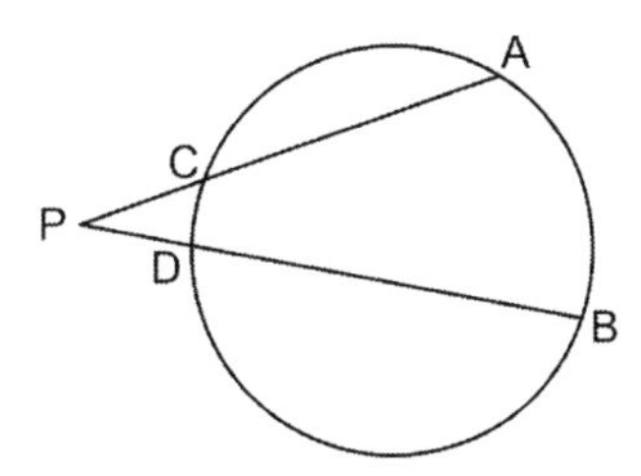

Then $\angle APB$ is half the difference of the angular sizes of arcs $\widehat{AB}$ and $\widehat{CD}$.

6.1 Example Questions

Problem 6.1 Consider the circle given by the equation $x^2 - 6x + y^2 + 4y = 0$.

(a) Find the center and radius of the circle by completing the square.

(b) What is the area and circumference of the circle?

Problem 6.2 Suppose you have a circle $(x-1)^2 + (y-1)^2 = 4$ and a line $x + y = 4$.

(a) The line and circle intersect at two points A, B, find them.

(b) Verify that the perpendicular bisector of $\overline{AB}$ goes through the center of the circle.

Problem 6.3 Let $\angle APB$ be an inscribed angle on a circle with center O. Prove that $\angle APB$ is half the angular size of arc $\widehat{AB}$ if:

(a) O lies on $\angle APB$.

(b) O lies inside $\angle APB$.

Problem 6.4 Prove that if two chords AC, BD intersect outside a circle at point P then the measure of $\angle APB$ is half the difference of the angular sizes of $\widehat{AB}, \widehat{CD}$.

Problem 6.5 Suppose $\widehat{AB}$ and $\widehat{CD}$ are arcs each with angular size 50° and if rays $\overrightarrow{BA}, \overrightarrow{DC}$ are extended to intersect at a point E (so A is on $\overline{BE}$ and on $\overline{DE}$), $\angle AEC = 50^\circ$. Find the angular size of arc $\widehat{BD}$. Express your answer in degrees, rounded to the nearest tenth if necessary.

Problem 6.6 Points A, B, C, and D are on a circle forming quadrilateral $ABCD$. If $\angle A = 2x$, $\angle B = 3x$, $\angle C = 5y$, and $\angle D = 3y$, what are x and y?

Problem 6.7 Points B and C are on a circle. A is such that $\overline{AB}$ and $\overline{AC}$ are both tangent to the circle. Prove that $AB = AC$.

Problem 6.8 A circle of radius AB is tangent to $\overline{BC}$ which has length $2\sqrt{5}+2$. $\overline{AC}$ intersects the circle at point D so that $CD = 4$ as shown in the diagram below.

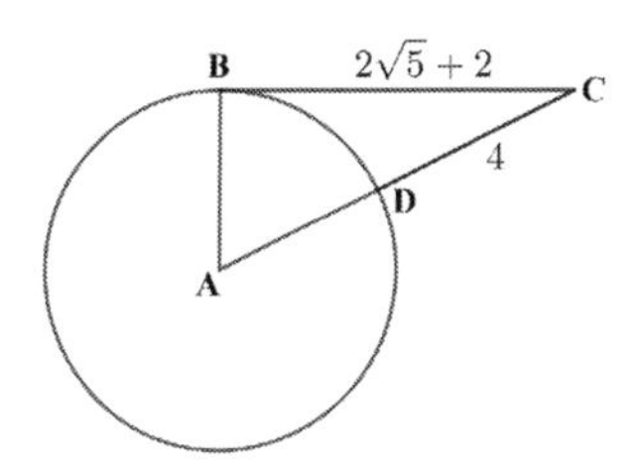

What is the area of the circle?

Problem 6.9 Consider the parallel lines $y = x + 4$ and $y = x - 4$. What is the equation of the largest circle with center $(0,0)$ that is tangent to both lines?

Problem 6.10 Find all k such that $y = kx$ is tangent to the circle $(x-4)^2 + (y-2)^2 = 4$.

6.2 Quick Response Questions

Problem 6.11 If four circles are drawn in a plane, what is the maximum number of points that belong to at least two of the circles?

Problem 6.12 Find the circumference of a circle with area 16π. Use $\pi = 3.14$. Round to the nearest tenth if necessary.

Problem 6.13 Find the area of a circle with circumference 16π. Use $\pi = 3.14$. Round to the nearest tenth if necessary.

Problem 6.14 Points A and B lie on a circle centered at point O with radius length 10 in. Suppose that the area measure of the sector formed by $\angle AOB$ is 20π in^2. How many degrees is $\angle AOB$?

Problem 6.15 Three circles, of radius 1, 2, and 3 are stacked vertically (largest to smallest) so that each circle is tangent to the next. How far away is the center of the circle of radius 3 from the center of the circle of radius 1?

Problem 6.16 A dog is tied to a leash of with a length of 10 feet. He tries to run around a nearby tree with diameter 3 feet, but gets stuck along the way (as the leash is too short). Using the picture below, which portion of the tree is where the dog gets stuck?

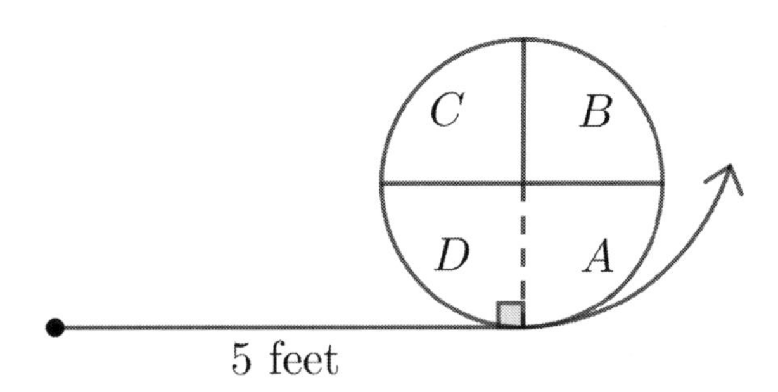

Problem 6.17 Two tangent circles are drawn with centers A and B in isosceles right triangle $\triangle ABC$.

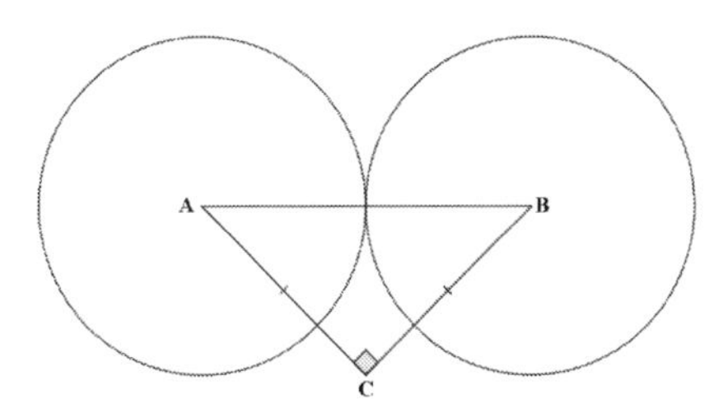

Using the approximation $\pi \approx \frac{22}{7}$ the area inside the triangle but outside the two circles is $\frac{P}{Q}$ of the entire triangle. What is $P+Q$?

Problem 6.18 Consider the sector shown below with a radius of 6.

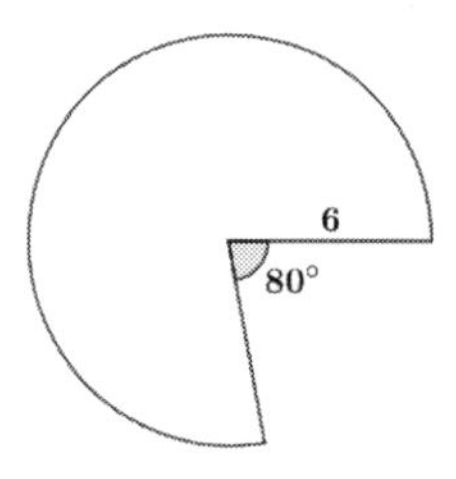

For this problem assume $\pi = 3$ and calculate the perimeter of the sector. Round your answer to the nearest integer if necessary.

Problem 6.19 In the following diagram $\angle ABC = 42^\circ$. What is the measure of $\angle AOC$?

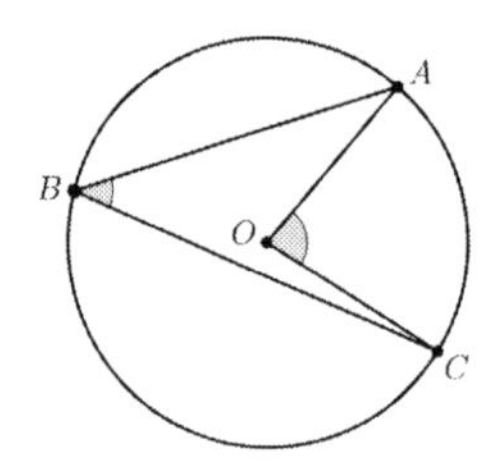

Problem 6.20 Suppose O is the center of a unit circle. Let $ABCO$ be a rhombus with A, B, C on the circle. $[ABCO] = \frac{\sqrt{P}}{Q}$. What is $P+Q$?

6.3 Practice Questions

Problem 6.21 A circle with center $(3,4)$ contains the point $(5,-2)$. Express the equation of this circle in the form $x^2+y^2=Ax+By+C$.

Problem 6.22 A circle $(x-1)^2+(y-1)^2=4$ and a line $x+y=4$ intersect at points A and B. If C is the center of the circle, find the area of $\triangle ABC$.

Problem 6.23 Let $\angle APB$ be an inscribed angle on a circle with center O. Prove that $\angle APB$ is half the angular measure of arc $\widehat{AB}$ if O lies outside $\angle APB$.

Problem 6.24 Prove that if two chords AC,BD intersect inside a circle at point P then the measure of $\angle APB$ is half the sum of the angular sizes of $\widehat{AB},\widehat{CD}$.

Problem 6.25 In the following diagram $\angle AOD=115°$ and $\angle BOC=23°$. What is the measure of angle $\angle BPC$?

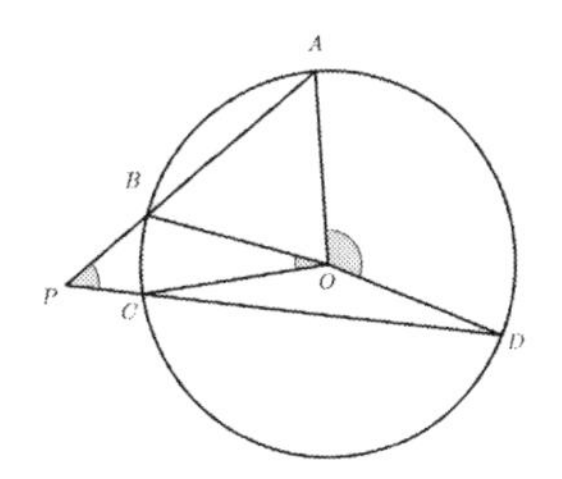

Problem 6.26 Suppose a trapezoid $ABCD$ with base $\overline{AD}$ is inscribed in a circle. If $AB = 4$, what is CD?

Problem 6.27 In the two diagrams given below, the lengths of segments $\overline{AB}$ are both equal to 2. Further suppose the inside circle in the second diagram has radius 1. Which shaded region has larger area?

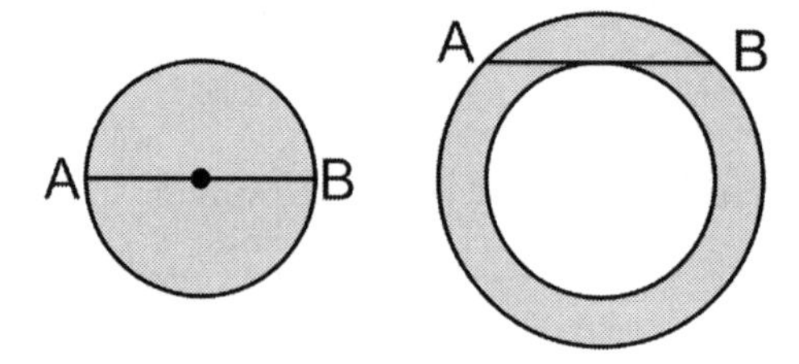

Problem 6.28 Two circles, C_1 and C_2, of radii 4 and 9 respectively, are tangent. Both circles are tangent to the line $y = 0$. The line $x = a$ is tangent to C_1 but does not intersect C_2 while the line $x = b$ is tangent to C_2 but does not intersect C_1. What is $|b - a|$? Round your answer to the nearest tenth if necessary.

Problem 6.29 Find the equation for the line with slope 1 that is tangent to the upper half of the circle $(x-3)^2 + (y-1)^2 = 2$.

Problem 6.30 Let $\mathscr{C}_1$ and $\mathscr{C}_2$ be circles defined by

$$(x-10)^2+y^2=36$$

and

$$(x+15)^2+y^2=81,$$

respectively. What is the length of the shortest line segment $\overline{PQ}$ that is tangent to $\mathscr{C}_1$ at P and to $\mathscr{C}_2$ at Q?

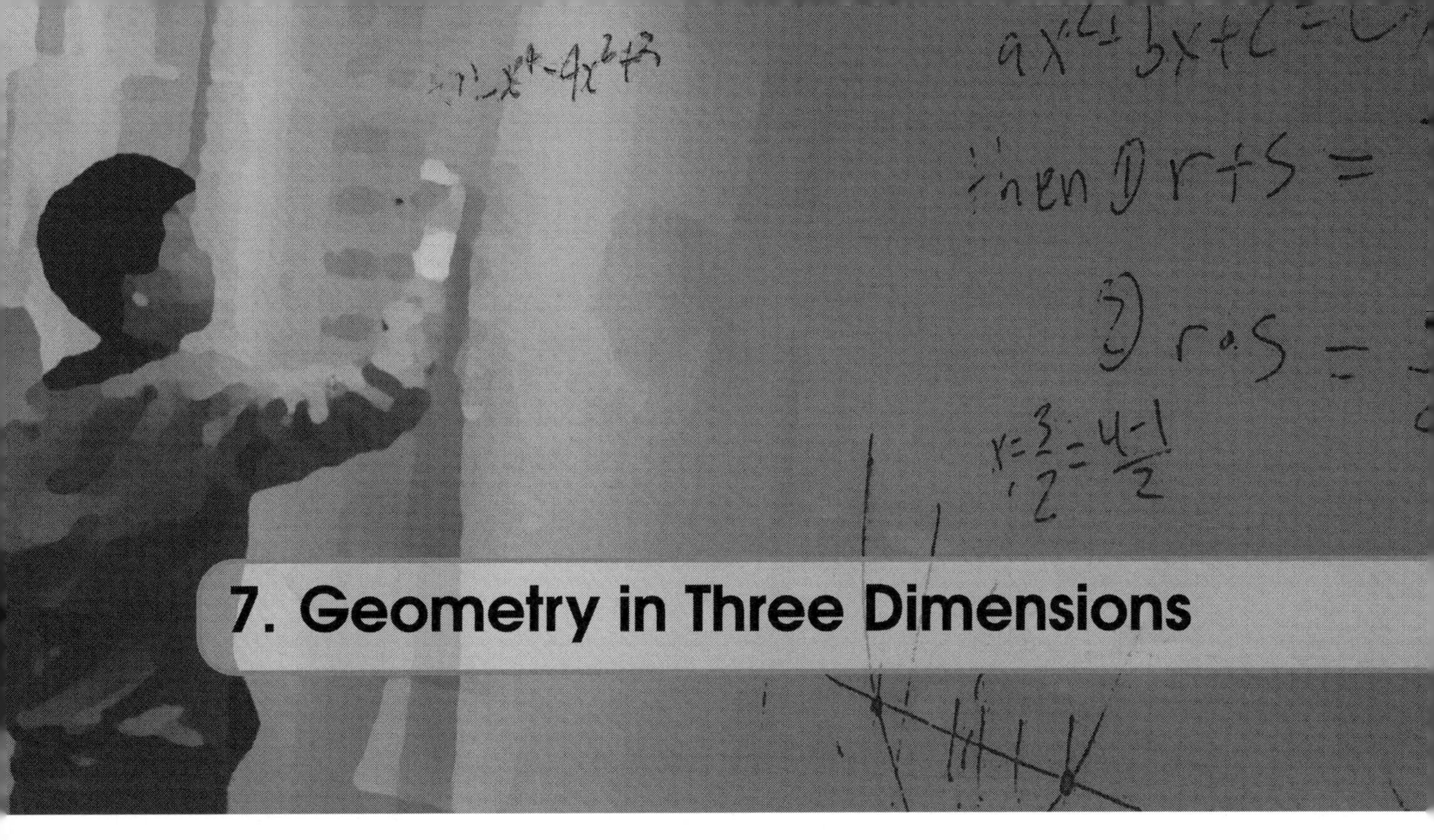

7. Geometry in Three Dimensions

Introduction

- Throughout the course, we have been doing what is referred to as *plane geometry*. That is geometry that takes place in a (two-dimensional) plane.
- Today we will study objects in three-dimensional. There are different ways of thinking of the three dimensions. Some common ways include:
    - length, width, and height.
    - up/down, left/right, forward/backward.
    - x-axis, y-axis, z-axis.
- We will call the entirety of the three-dimensions we work with *space*.
- Points in space can be thought of as using x, y, z coordinates, written (x, y, z).
- Three-dimensional objects are often called *solids*.
- Two of the simplest solids are *balls* and *cubes*.

Volume and Surface Area

- The amount of space a solid takes up is called its *volume*.
- For example, the amount of water a solid can hold is given by its volume.
- The amount of area on the outside of the solid is called its *surface area*.
- For example, the amount of paint needed to cover the outside of a solid is given by its surface area.

Review of Lower Dimensional Space

- *Points* are 0-dimensional objects.

- *Lines* and *line segments* are 1-dimensional objects. Remember lines are go on forever, but line segments are finite and have a finite *length.*
- *Planes* are 2-dimensional objects (think of a piece of paper extended in all directions). We called finite pieces of a plane *shapes* and shapes have a finite *area.*

Summary of Basic Results

- Given two distinct lines, there are three possibilities: (i) they are parallel, (ii) they intersect, or (iii) they are *skew*.
- Given two distinct planes, there are two possibilities: (i) they are parallel, or (ii) they intersect (and their intersection is a line).
- We have analogues of the Pythagorean theorem and Distance Formula in three dimensions: The distance from $(0,0,0)$ to (a,b,c) is $\sqrt{a^2+b^2+c^2}$. The distance from $A=(x_A,y_A,z_A)$ to $B=(x_B,y_B,z_B)$ is $\sqrt{(x_A-x_B)^2+(y_A-y_B)^2+(z_A-z_B)^2}$.

7.1 Example Questions

Problem 7.1 In two-dimensional geometry (inside a plane), two different lines are either parallel or they intersect.

(a) Come up with a definition of 'parallel' that makes sense in three-dimensions.

(b) Argue that two different lines in space can be non-intersecting and not parallel at the same time. (Such lines are called *skew* lines.)

Problem 7.2 The two equations $x+y+z=1$ and $x-y=0$ each give an equation of a plane.

(a) Graph the intersection of these planes when $x=0$, when $y=0$, and when $z=0$.

(b) The two planes intersect to form a line. Show that the expression $x = y = \frac{z-1}{-2}$ determines this line.

Problem 7.3 Distance Formula in Three-dimensions: Distance from $(0,0,0)$ to (x,y,z).

(a) If $A = (0,0,0)$, $B = (x,y,z)$, and $C = (x,y,0)$ argue that $\triangle ACB$ is a right triangle.

(b) By first calculating AC and BC, use the Pythagorean theorem to show that $AB = \sqrt{x^2+y^2+z^2}$.

Problem 7.4 You have a cube with side length 1. Label the vertices of the square $ABCD-EFGH$ where $ABCD$ forms the "bottom" square and $EFGH$ forms the "upper" square as in the diagram below.

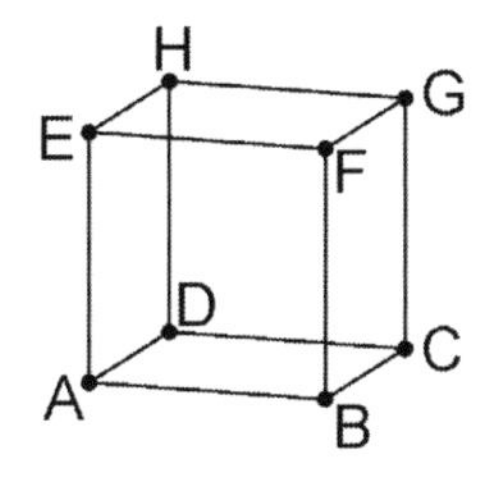

Find the distance from A to B, from A to F, and from A to G.

Problem 7.5 How many vertices, edges, and faces does a cube have? Describe the faces.

Problem 7.6 Euler's Theorem

(a) Euler's Theorem states that $F - E + V$ is a constant for all polyhedron. Here F is the number of faces, V the number of vertices, and E the number of edges. Using cubes as an example, what is this constant?

(b) Verify Euler's Theorem for a regular dodecahedron which is formed by 12 regular pentagons and shown below.

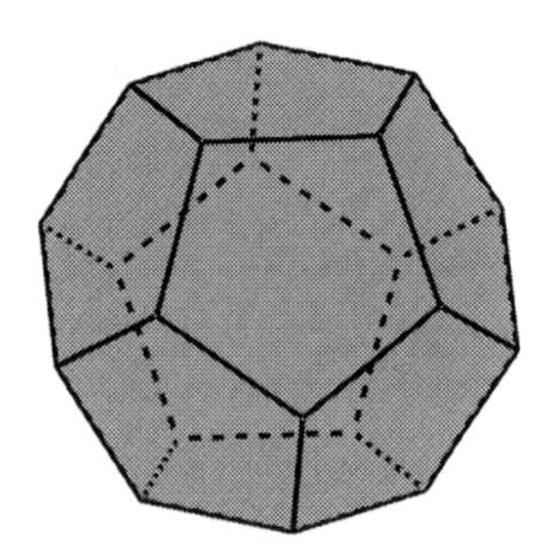

Problem 7.7 You have a box (rectangular prism) that is 2 feet long, 1 foot wide, and has a height of 6 inches.

(a) How much space is inside the box? That is, what is the volume of the box?

(b) What is the surface area of the box?

(c) You want to double the volume of the box by changing one of the dimensions of the box. What are the possible new surface areas (measured in square feet)?

Problem 7.8 A cube is increased to form a new cube so that the surface area of the new cube is 4 times that of the original cube. By what factor is the side length increased? What about the volume of the cube?

Problem 7.9 Consider a ball (or sphere) with radius 6.

(a) Find the volume of the ball using the formula $\frac{4}{3}\pi r^3$.

(b) Find the surface area of the ball using the formula $4\pi r^2$.

Problem 7.10 If a solid ball with radius 6 is cut in half, find the volume and surface area of the half-ball.

7.2 Quick Response Questions

Problem 7.11 What is the distance between the points $(2,2,4)$ and $(6,2,1)$?

Problem 7.12 Consider the diagram below.

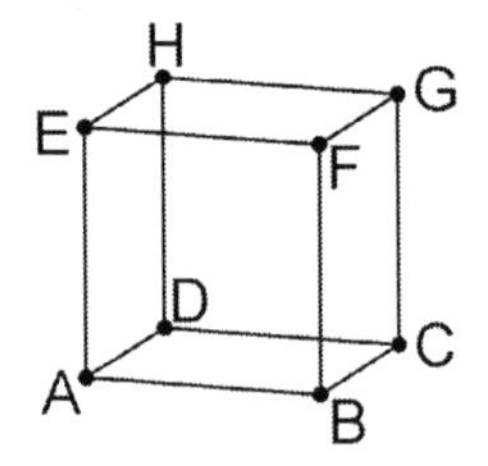

Which of the following does not intersect $\overline{AB}$?

(A) $\overline{BC}$
(B) $\overline{EF}$
(C) $\overline{EA}$
(D) $\overline{AD}$

Problem 7.13 Consider the diagram below.

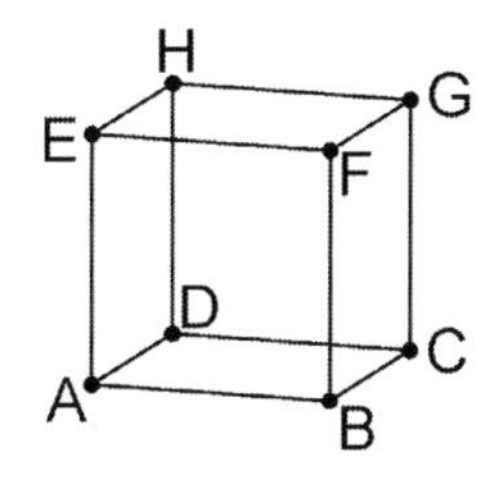

Which of the following is not parallel to $\overline{AB}$?

(A) $\overline{BC}$
(B) $\overline{EF}$
(C) $\overline{HG}$
(D) $\overline{DC}$

Problem 7.14 Consider the diagram below.

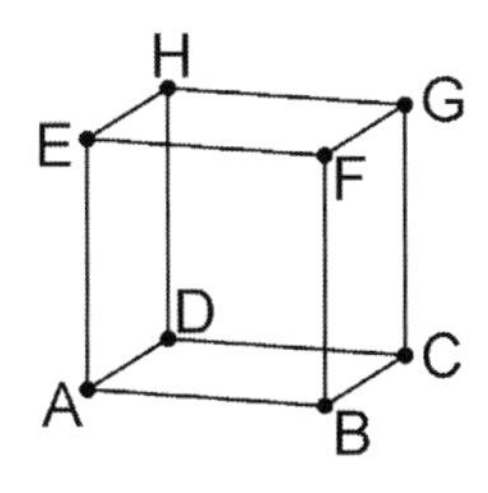

Which of the following is not skew to $\overline{AB}$

(A) $\overline{CG}$
(B) $\overline{HG}$
(C) $\overline{EH}$
(D) $\overline{FG}$

Problem 7.15 What is the volume of a sphere with radius 4? Round your answer to the nearest integer.

Problem 7.16 What is the surface area of a sphere with radius 4? Round your answer to the nearest integer.

Problem 7.17 Consider a rectangular prism with dimensions 6, 10, and 15, and a cube with side length 10. Which of the following statements is true?

(A) They have the same volume and same surface area
(B) They have the same surface area but different volume
(C) They have the same volume but different surface area
(D) They have different volume and surface area

Problem 7.18 A rectangular prism has face with area 35 another face with area 42. If all dimensions of the prism are of integer length bigger than 1, what is the surface area of the rectangular prism?

Problem 7.19 Let $A = (-2,0,0)$, $B = (2,0,0)$, $C = (0,2,0)$, $D = (0,-1,0)$, $E = (0,0,1)$ and $F = (0,0,-1)$ be the vertices of an octahedron. What is the length of its longest diagonal?

Problem 7.20 A polyhedron has 12 vertices and 18 edges. How many faces does it have?

7.3 Practice Questions

Problem 7.21 Yes or No. Explain your answers! All the questions take place in (three-dimensional) space.

(a) Given two points, there is a (unique) line containing the points.

(b) Given three different points, there is a (unique) plane containing all three points.

(c) Given a line and a point not on the line, there is a (unique) line containing the point that is parallel to the first line.

Problem 7.22 The two equations for the planes $x+y+z=1$ and $x-y=0$ intersect to form the line determined by the equation $x=y=\dfrac{z-1}{-2}$.

What expression determines the line formed by the intersection of the planes $x+y=2$ and $y-z=1$?

Problem 7.23 A rectangular prism has length 3 inches and width 4 inches. If the diagonal connecting opposite vertices inside the prism has length 13, what is the height of the prism?

Problem 7.24 How many total pairs of parallel edges does a cube contain?

Problem 7.25 A tetrahedron is a polyhedron formed using only equilateral triangles for the faces, as shown below.

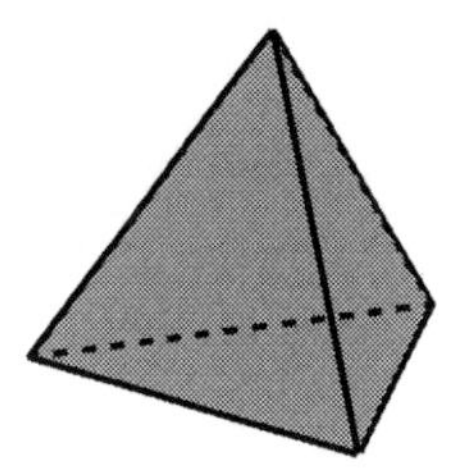

How many vertices, edges, and faces does a tetrahedron have?

Problem 7.26 Verify Euler's Theorem for a regular octahedron which is formed by 8 equilateral triangles and shown below.

Problem 7.27 Suppose a rectangular block of wood with dimensions 5 cm $\times 6$ cm $\times 5$ cm costs \$50. In dollars, what is the fair price for a rectangular block of the same type of wood with dimensions 15 cm by 30 cm by 40 cm if the price is determined solely by volume?

Problem 7.28 A cube is increased to form a new cube so that the surface area of the new cube is 64 times that of the original cube. By what factor is the volume of the cube increased?

Problem 7.29 Compare the volume of a ball with radius 4 to the combined volumes of two balls of radius 3.

Problem 7.30 Compare the surface area of a ball with radius 4 to the combined surface areas of two balls of radius 3.

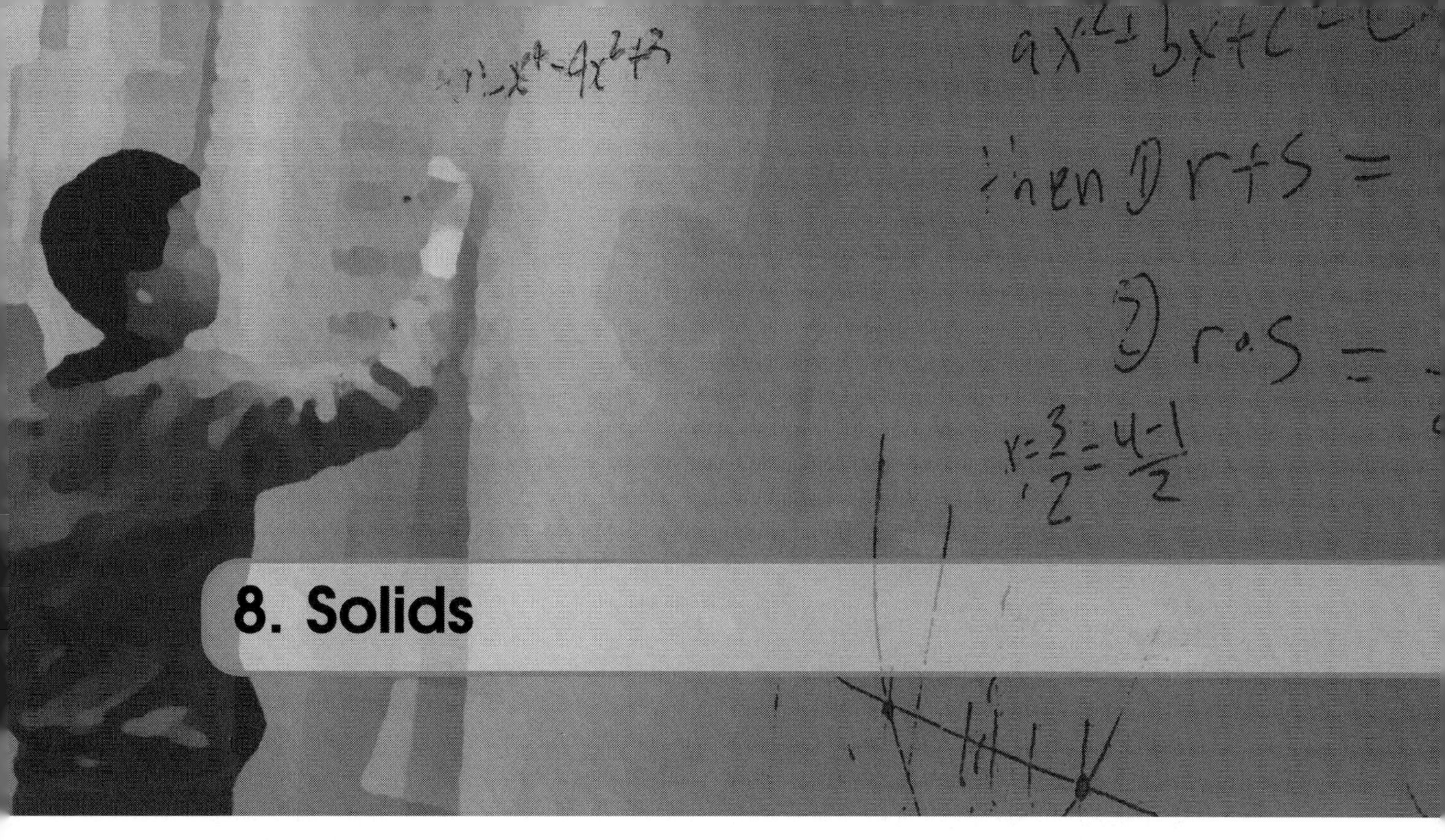

8. Solids

Solids Review

- **Sphere**: The collection of points (in three dimensions) of equal distance (called the *radius* from a *center*).
  - A sphere with radius r has volume $\frac{4}{3}\pi r^3$.
  - A sphere with radius r has surface area $4\pi r^2$.
- **Cube**: The 3-D version of a square.
  - A cube with side length s has volume s^3.
  - A cube with side length s has surface area $6s^2$.
  - A cube has 8 vertices, 12 edges, and 6 faces. The 6 faces are all squares.
- **Rectangular Prism**: A "box".
  - A rectangular prism with sides l, w, h (often 'length', 'width', 'height') has volume lwh.
  - A rectangular prism with sides l, w, h has surface area $2(lw + wh + lh)$.
  - Like a cube, a rectangular prism has 8 vertices, 12 edges, and 6 faces. The 6 faces are all rectangles.

Examples of Common Solids

- **Cylinder**: A "can".
  - A cylinder with height h and radius r has volume $\pi r^2 h$.
  - A cylinder with height h and radius r has surface area $2\pi r^2 + 2\pi rh$.
- **Square Right Pyramid**: The standard "pyramid from Egypt" solid, with a square *base* and *apex* (or top point) that is centered above the square.

- A square right pyramid with height h with square base of side length s has volume $\frac{1}{3}s^2h$.
- A square right pyramid with height h with square base of side length s has surface area $s^2 + 2sL$, where $L = \sqrt{h^2 + s^2/4}$ (L is called the *slant height*).
- A square right pyramid has 5 vertices, 8 edges, and 5 faces. The 4 side faces are all triangles.

- **Right Cone**: The standard "ice cream cone" solid, with a circular *base* and *apex* that is centered above the circle.
  - A right with height h with square base of radius r has volume $\frac{1}{3}\pi r^2 h$.
  - A square right pyramid with height h with square base of side length s has surface area $\pi r^2 + \pi rL$, where $L = \sqrt{h^2 + r^2}$ (L is called the *lateral height*).

8.1 Example Questions

Problem 8.1 Cutting a rectangular prism in half along the diagonal will form a triangular prism as shown below.

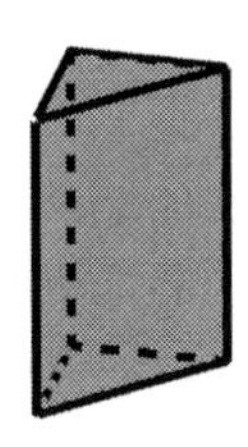

The triangular prism shown has a height of 2 cm and the base is an isosceles right triangle with legs 1 cm long.

(a) Find the volume of the prism. Explain how to extend your method to a general formula.

(b) Find the surface area of the prism. Does your method extend to a general formula?

Problem 8.2 Consider the cylinder (a circular prism) shown below with height 3 and radius 1.

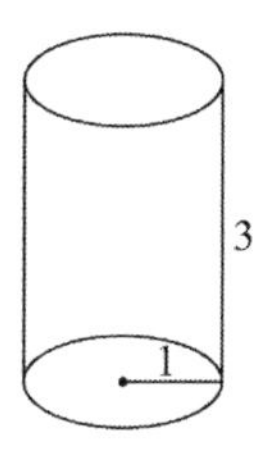

(a) What is the volume of the cylinder?

(b) What is the surface area of the cylinder?

Problem 8.3 Consider the shapes of a pyramid or a cone. They fit inside a rectangular prism or a cylinder as shown below:

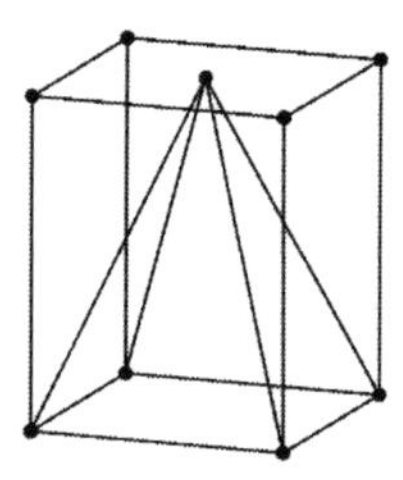

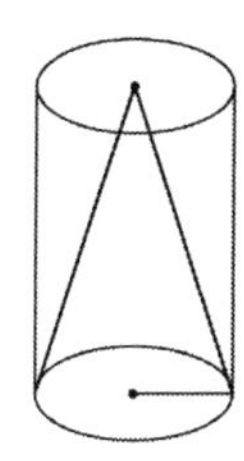

In general, the volumes of pyramids and cones are $\frac{1}{3}$ the volume of the full shape.

(a) Find the volume of a square right pyramid with base of side length 4 and height 5.

(b) Find the volume of a cone with radius 3 and height 4.

(c) Find the volume of a triangular pyramid with height 3 and an equilateral triangle base with side length 3.

Problem 8.4 Find the surface area of a square right pyramid with base of side length 4 and height 5.

Problem 8.5 Find the surface area of a cone with radius 3 and height 4.

Problem 8.6 Suppose you have an ice cream cone with radius 2 inches and height 4 inches. The cone starts full of ice cream (but there is not ice cream outside the cone). After you've eaten some ice cream and some of the cone you are left with a cone with a radius and a height of 2 inches. What fraction of the ice cream have you eaten?

Problem 8.7 Inscribing Spheres and Cubes

(a) Find the volume of the largest sphere that fits in a cube of volume 1. (That is, inscribe a sphere inside the cube.)

(b) Find the volume of the smallest sphere that holds a cube of volume 1. (That is, circumscribe a sphere outside the cube.)

Problem 8.8 Consider a unit cube $ABCD-EFGH$. Cut the cube through $\triangle AFH$ into two smaller solids.

Describe these two solids. How many faces, edges, and vertices does each have? What is the volume of each?

Problem 8.9 Consider a unit cube $ABCD-EFGH$. Cut the cube through $\triangle AFH$ into two smaller solids.

What is the surface area of the remaining two solids?

Problem 8.10 An obtuse triangle with dimensions 9, 10, and 17 is rotated about the smallest side so that it creates a three-dimensional solid shown below.

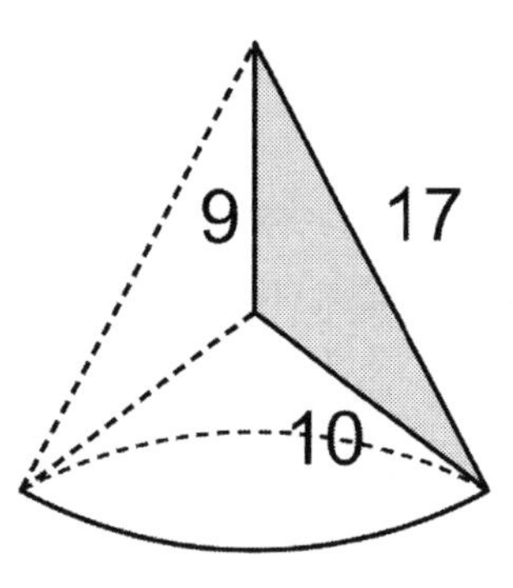

Determine the volume of the solid.

8.2 Quick Response Questions

Problem 8.11 A triangular right prism has height 13 and the triangular base has sides 8, 15, and 17. What is the volume of the prism? Round your answer to the nearest integer if necessary.

Problem 8.12 A triangular right prism has height 13 and the triangular base has sides 8, 15, and 17. What is the surface area of the prism? Round your answer to the nearest integer if necessary.

Problem 8.13 A cylinder has a base with area 9π. If its volume is 90π, what is its height? Round your answer to the nearest integer if necessary.

Problem 8.14 A cylinder has a base with area 9π. If its surface area is 90π, what is its height? Round your answer to the nearest integer if necessary.

Problem 8.15 What is the volume of a square pyramid with base side length 6 and height 4? Round your answer to the nearest integer if necessary.

Problem 8.16 What is the surface area of a square pyramid with base side length 6 and height 4? Round your answer to the nearest integer if necessary.

Problem 8.17 David creates a solid by starting with a cube of side length 4 cm and adding a square pyramid with height 2 cm to the top, so that the base of the pyramid is is one of the faces of the original cube.

How many faces does this solid have?

Problem 8.18 David creates a solid by starting with a cube of side length 4 cm and adding a square pyramid with height 2 cm to the top, so that the base of the pyramid is is one of the faces of the original cube.

How many edges does this solid have?

Problem 8.19 David creates a solid by starting with a cube of side length 4 cm and adding a square pyramid with height 2 cm to the top, so that the base of the pyramid is is one of the faces of the original cube.

How many vertices does this solid have?

Problem 8.20 David creates a solid by starting with a cube of side length 4 cm and adding a square pyramid with height 2 cm to the top, so that the base of the pyramid is is one of the faces of the original cube.

What is the volume of this solid? Round your answer to the nearest cubic centimeter if necessary.

8.3 Practice Questions

Problem 8.21 The triangular prism shown below has all edges of length 1.

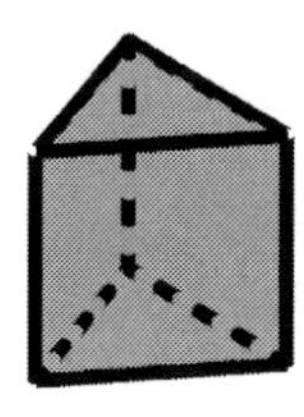

What is its volume and surface area?

Problem 8.22 Suppose you want to build a can of soda to hold 12π cubic inches of soda. If you want the diameter of the can to be 8 inches, how tall should you make the can?

Problem 8.23 A regular tetrahedron is formed using 4 equilateral triangles. If each equilateral triangle has a side length of 2, what is the volume of the tetrahedron? The diagram shown below should be helpful in determining the height.

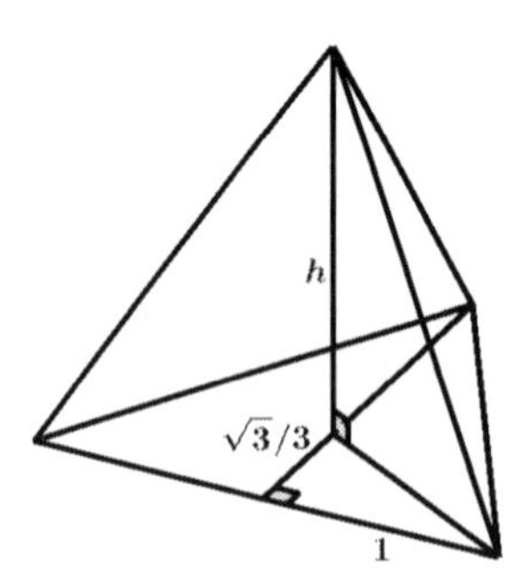

Problem 8.24 A regular tetrahedron is formed using 4 equilateral triangles. If each equilateral triangle has a side length of a, what is the surface area of the tetrahedron?

Problem 8.25 The circular sector with radius 8 shown below is folded together to make a cone.

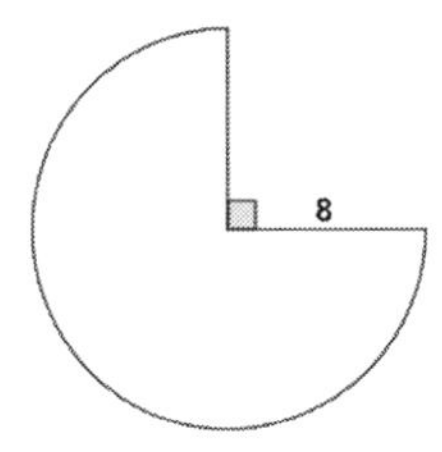

What is the area of the base of this cone?

Problem 8.26 Darren has a wooden cone with radius 3 and height 4. Darren wants to cut the cone horizontally to create two pieces (one of which will be a smaller cone similar to the original) so that each of the pieces has the same volume.

At what height should Darren cut the cone? (That is, what is the height of the smaller cone similar to the original.)

Problem 8.27 A cone and a cylinder both have the same height and the same volume. What is the ratio of the radius of the cone to the radius of the cylinder?

Problem 8.28 Consider a unit cube $ABCD-EFGH$. Form a tetrahedron by connecting the vertices A, C, F, and H.

Find the lengths of all the edges of the tetrahedron.

Problem 8.29 Consider a unit cube $ABCD-EFGH$. Form a tetrahedron by connecting the vertices A, C, F, and H.

Find the volume of the tetrahedron.

Problem 8.30 An obtuse triangle with dimensions 9, 10, and 17 is rotated about the smallest side so that it creates a three-dimensional solid shown below. Determine the surface area of the solid.

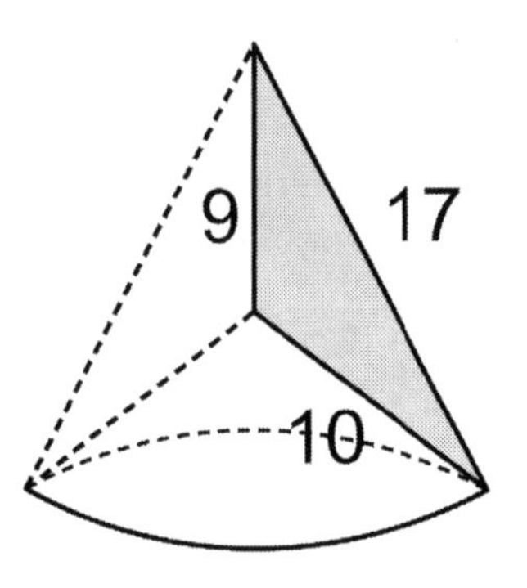

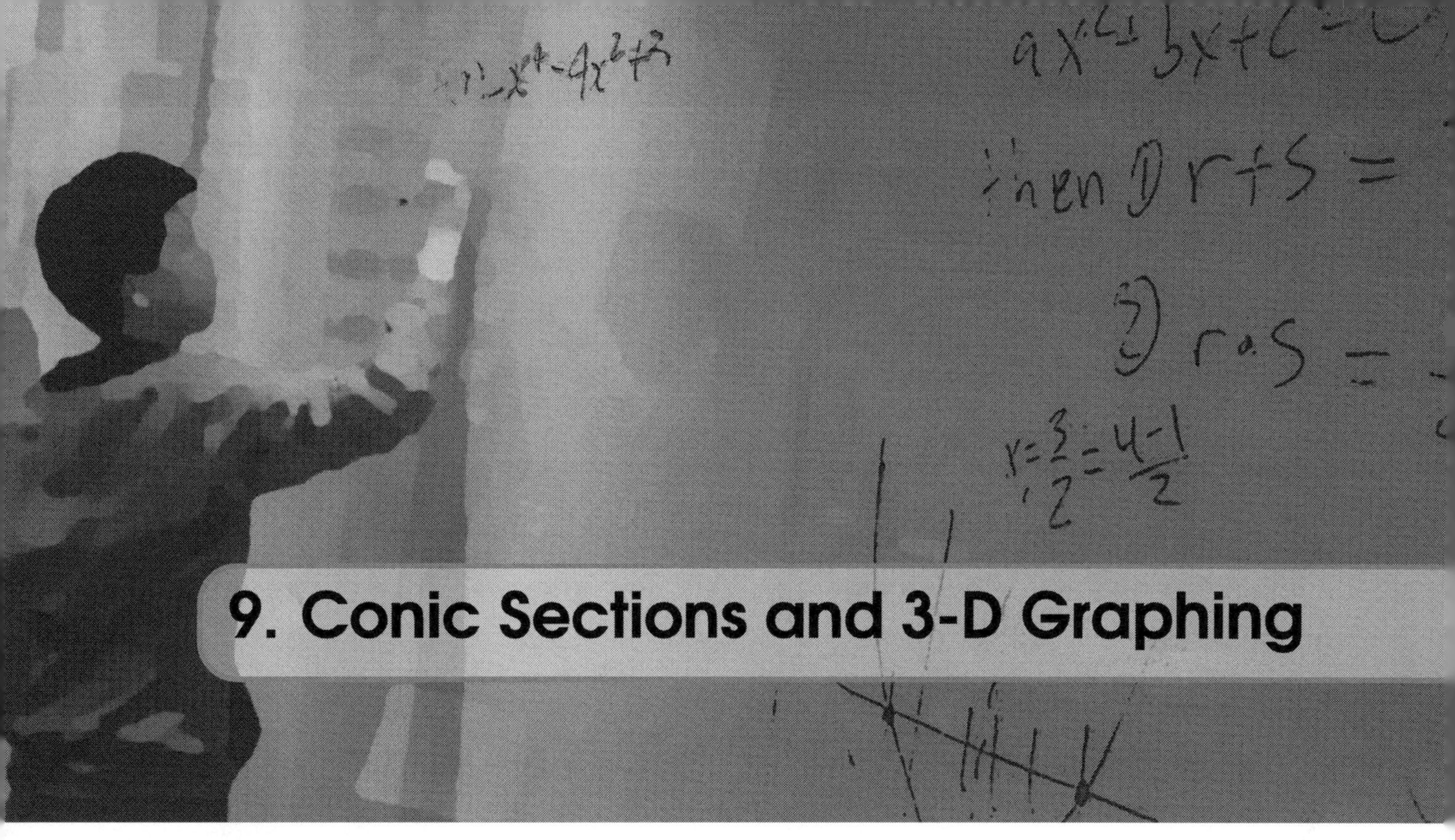

9. Conic Sections and 3-D Graphing

Conic Sections

- Conic sections are formed by the intersection of a plane with a cone.
- The four types of conic sections are:
  - Circles
  - Ellipses
  - Parabolas
  - Hyperbolas
- Equations centered at the origin are:
  - Circle: $x^2 + y^2 = r^2$
  - Ellipses: $\frac{x^2}{a^2} + \frac{y^2}{b^2} = 1$
  - Parabolas $y = kx^2$ or $x = ky^2$
  - Hyperbolas $\frac{x^2}{a^2} - \frac{y^2}{b^2} = 1$ or $\frac{y^2}{a^2} - \frac{x^2}{b^2} = 1$
- Translations:
  - Replacing x by $x - h$ in an equation shifts the graph h units right.
  - Replacing y by $y - k$ in an equation shifts the graph k units up.

9.1 Example Questions

Problem 9.1 Consider the graph of the equation $x^2 + y^2 = z^2$ in three dimensions.

(a) What does the graph look like if $x = 0$? What about if $y = 0$?

(b) Describe the graph if $z = c$ for a constant c.

(c) Based on the graphs from parts (a) and (b) (which are often called cross-sections), what geometric object is graphed by the equation?

Problem 9.2 Find the volume of the solid region formed by the inequality $x^2 + y^2 \geq 4z^2$ when $0 \leq z \leq 9$.

Problem 9.3 Consider the double cone $x^2 + y^2 = z^2$.

We saw that cross sections when z was a constant were circles. Find equations and describe the graph for the cross sections when x is constant or when y is constant.

Problem 9.4 Consider the double cone $x^2 + y^2 = z^2$.

Find the equation and describe the graph of the cross section we get when $z = y + 1$.

Problem 9.5 Consider the double cone $x^2 + y^2 = z^2$.

Find the equation and describe the graph of the cross section we get when $z = \dfrac{y}{2} + 1$.

Problem 9.6 Consider the double cone $x^2+y^2=z^2$ and the plane $z=my+1$. Describe the values of m which produce a circle, ellipse, parabola, and hyperbola.

Problem 9.7 Graph the ellipse with equation $\frac{x^2}{4}+\frac{y^2}{9}=1$.

Problem 9.8 Consider the hyperbola with equation $\frac{x^2}{4}-\frac{y^2}{9}=1$.

(a) Argue that when x and y are large the equation of the hyperbola can be approximated by $y=\pm\frac{3}{2}x$.

(b) Use the approximation from part (a) to graph the hyperbola.

Problem 9.9 The graph of the equation $x^2+y^2=z$ is referred to as a paraboloid. What are the cross sections when x, y, or z are constant? Describe the graph.

Problem 9.10 The graph of the equation $x^2+y^2=z^2+1$ is referred to as a one-sheeted hyperboloid. What are the cross sections when x, y, or z are constant? Describe the graph.

9.2 Quick Response Questions

Problem 9.11 The equation $\frac{(x-3)^2}{4}+(y+1)^2=1$ gives what type of conic?

(A) A circle
(B) A parabola
(C) An ellipse
(D) A hyperbola

Problem 9.12 The equation $x^2+5y^2=(2y+3)^2$ gives what type of conic?

(A) A circle
(B) A parabola
(C) An ellipse
(D) A hyperbola

Problem 9.13 The equation $x^2-2y^2=-8$ gives what type of conic?

(A) A circle
(B) A parabola
(C) An ellipse
(D) A hyperbola

Problem 9.14 The equation $x^2-2y^2=(x+3)^2$ gives what type of conic?

(A) A circle
(B) A parabola
(C) An ellipse
(D) A hyperbola

Problem 9.15 A right circular cone has a radius of 6 inches. Its volume and surface area are equal. What is the height of the cone in inches? Round your answer to the nearest tenth if necessary.

Problem 9.16 For graph of the surface $x^2+y^2=(z-2)^2$ creates the outside of a cone. The volume of the cone for $0 \le z \le 2$ can be written as $K \times \pi$ for a number K. What is K rounded to the nearest tenth?

Problem 9.17 Which of the following gives the cross section of $x^2+y^2=(z-2)^2$ when $x=2$?

(A) $y^2-z^2=4z$
(B) $y^2+z^2=4z$
(C) $z^2+y^2=4z$
(D) $z^2-y^2=4z$

Problem 9.18 Which of the following gives the cross section of $x^2+y^2=(z-2)^2$ when $z=y$?

(A) $4-x^2=4y$
(B) $4+x^2=4y$
(C) $4+x^2=2y$
(D) $4-x^2=2y$

Problem 9.19 Which of the following gives the cross section of $x^2+y^2=(z-2)^2$ when $2z=y$?

(A) $2x^2+y^2+4y=16$
(B) $2x^2+y^2+4y=8$
(C) $4x^2+3y^2+8y=16$
(D) $4x^2+3y^2+8y=8$

Problem 9.20 Is it true that the equation $x^2 - 4x + y^2 - 2y = z^2 - 5$ gives the equation of a cone?

9.3 Practice Questions

Problem 9.21 Consider the four planes: (i) $z = 1 - x - y$, (ii) $z = 1 + x - y$, (iii) $z = 1 - x + y$, and (iv) $z = 1 + x + y$.

(a) Using a single graph, draw the intersections of the four planes with the plane $z = 0$.

(b) Using a single graph, draw the intersections of the four planes with the plane $z = 0.5$.

(c) Using a single graph, draw the intersections of the four planes with the plane $z = 1$.

Problem 9.22 The four planes (i) $z = 1 - x - y$, (ii) $z = 1 + x - y$, (iii) $z = 1 - x + y$, and (iv) $z = 1 + x + y$ along with the plane $z = 0$ form the boundary of a square pyramid. Find the volume of this pyramid.

Problem 9.23 Consider the double cone $x^2 + z^2 = (y + 3)^2$. The intersection of this cone with the plane $y = 1$ forms a circle. What is the circumference of this circle?

Problem 9.24 Consider the double cone $x^2 + z^2 = (y + 3)^2$. The intersection of this cone with the plane $y = x$ forms a parabola. What is the vertex of this parabola? Remember that this vertex will have have an x, y, and z coordinate.

Problem 9.25 Consider the double cone $x^2+z^2=(y+3)^2$. The intersection of this cone with the plane $y=\frac{z}{4}$ forms an ellipse. Find and list 4 points on this ellipse.

Problem 9.26 Graph the equation $9x^2+25y^2=225$. What type of conic is this?

Problem 9.27 Graph the equation $-9x^2+25y^2=225$. What type of conic is this?

Problem 9.28 Find the intersection points of the two conic sections $x^2-y^2=6$ and $3x^2+y^2=30$.

Problem 9.29 The graph of the equation $\frac{x^2}{4}+y^2=z$ is referred to as an elliptic paraboloid. What are the cross sections when x, y, or z are constant? Describe the graph.

Problem 9.30 The graph of the equation $x^2+y^2=z^2-1$ is referred to as an two-sheeted hyperboloid. What are the cross sections when x, y, or z are constant? Describe the graph.

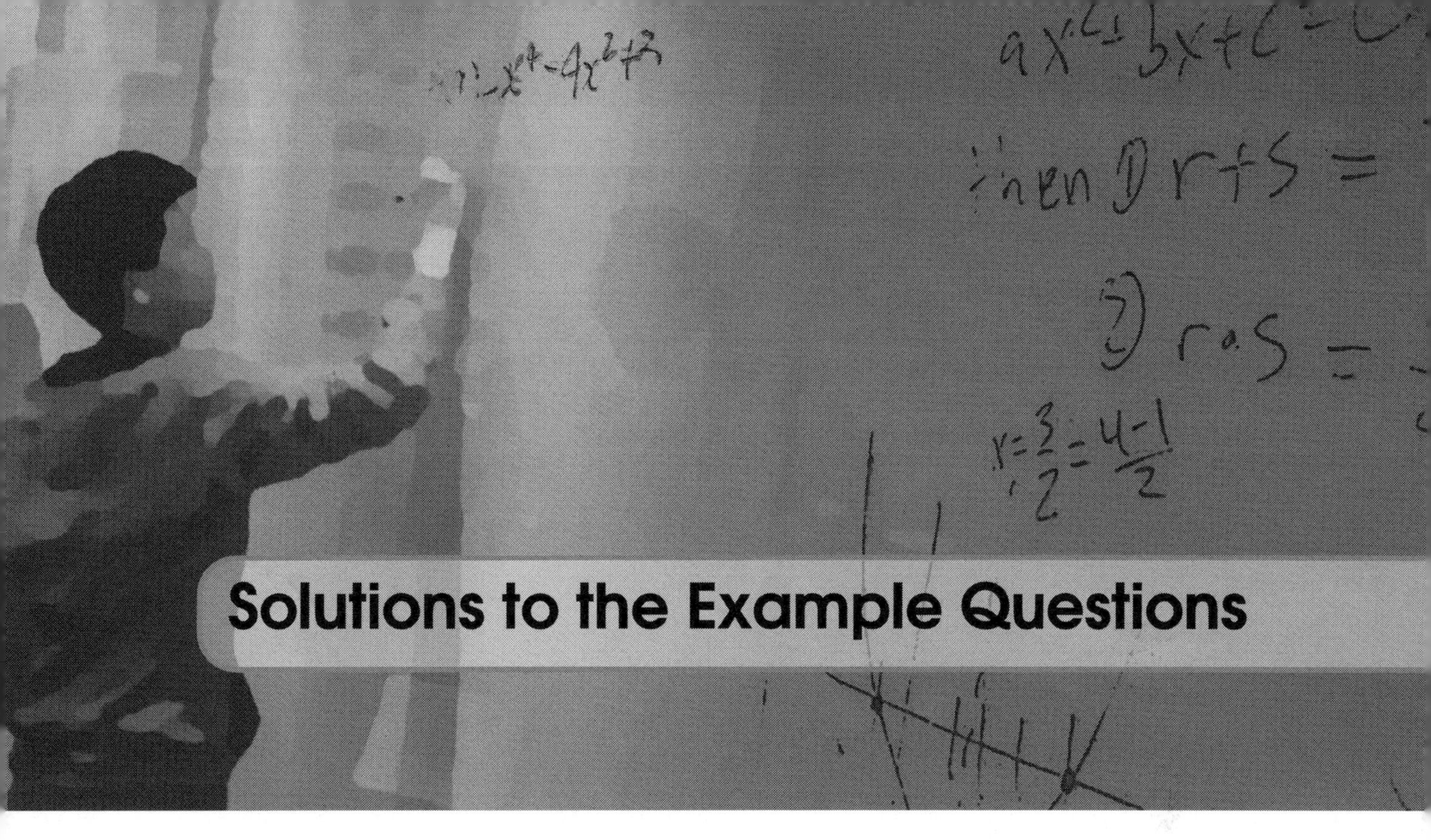

Solutions to the Example Questions

In the sections below you will find solutions to all of the Example Questions contained in this book.

Quick Response and Practice questions are meant to be used for homework, so their answers and solutions are not included. Teachers or math coaches may contact Areteem at info@areteem.org for answer keys and options for purchasing a Teachers' Edition of the course.

1 Solutions to Chapter 1 Examples

Problem 1.1 Let $A = (0,0)$, $B = (-2,1)$, and $C = (3,6)$.

(a) Plot the points A, B, and C on a coordinate plane.

Solution

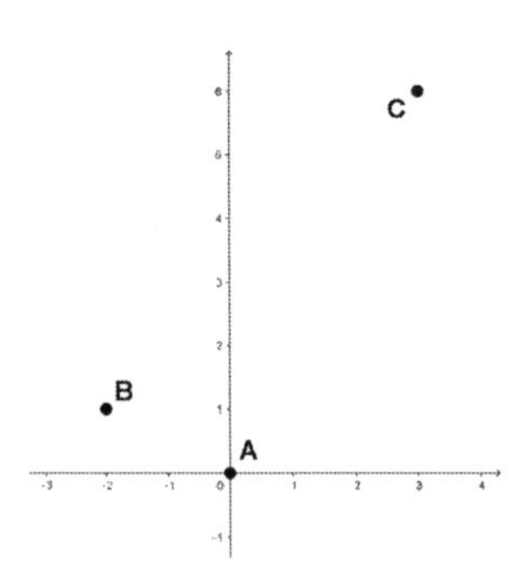

(b) Find the equation of the line containing points A and B.

Answer

$y = -\frac{1}{2}x$

Solution

The slope from A to B is $\frac{0-1}{0-(-2)} = \frac{-1}{2}$ so as the line contains the origin the equation is $y = -\frac{1}{2}x$.

(c) Find the equation of the line containing points B and C.

Answer

$y = x + 3$

Solution

The slope from B to C is $\frac{6-1}{3-(-2)} = 1$ so using point-slope form the equation is

$$y - 6 = x - 3 \text{ or } y = x + 3.$$

Problem 1.2 Prove that the equation for the line going through points A and B is given by

$$\frac{x-x_A}{x_A-x_B}=\frac{y-y_A}{y_A-y_B}.$$

Solution

Note that a point $P=(x,y)$ is on the same line as A and B if and only if the slope from P to A and A to B are the same. That is

$$\frac{y-y_A}{x-x_A}=\frac{y_A-y_B}{x_A-x_B} \text{ or (after rearranging) } \frac{x-x_A}{x_A-x_B}=\frac{y-y_A}{y_A-y_B}$$

as needed.

Problem 1.3 Let $A=(1,2)$ and $B=(3,y_B)$. What is y_B if $\overline{AB}$ is parallel the line $y=3x-4$?

Answer

8

Solution

The slope of the line $y=3x-4$ is 3. Hence we want the slope from A to B to be 3. Therefore

$$3=\frac{y_B-2}{3-1}\Rightarrow 6=y_B-2\Rightarrow y_B=8$$

so $y_B=8$.

Problem 1.4 Given a line segment $\overline{AB}$, the perpendicular bisector of $\overline{AB}$ is a line perpendicular to $\overline{AB}$ going through its midpoint.

If $A=(1,1)$ and $B=(5,3)$, find the perpendicular bisector of $\overline{AB}$.

Answer

$y=-2x+8$

Solution

First calculate the midpoint of $\overline{AB}$ which is

$$\left(\frac{1+5}{2}, \frac{1+3}{2}\right) = (3,2).$$

The slope from A to B is

$$\frac{3-1}{5-1} = \frac{1}{2}.$$

Therefore we want the line with slope -2 containing $(3,2)$, which, using point-slope form, is

$$y-2 = -2(x-3) \Rightarrow y = -2x+8.$$

Problem 1.5 Let $O = (0,0)$, $A = (-1,0)$, and $C = (1,0)$. If B is a point such that $\angle AOB = 30^\circ + \angle BOC$, what is $\angle AOB$?

Answer

105°

Solution

A, O, and C are all on the same line, so $\angle AOC = 180^\circ$. Therefore

$$180^\circ = \angle AOC = \angle AOB + \angle BOC = 30^\circ + \angle BOC + \angle BOC$$

so $2\angle BOC = 150^\circ$ and thus $\angle BOC = 75^\circ$. Hence $\angle AOB = 75^\circ + 30^\circ = 105^\circ$.

Problem 1.6 Consider the diagram below, where l and m are parallel but the drawing is not necessarily to scale.

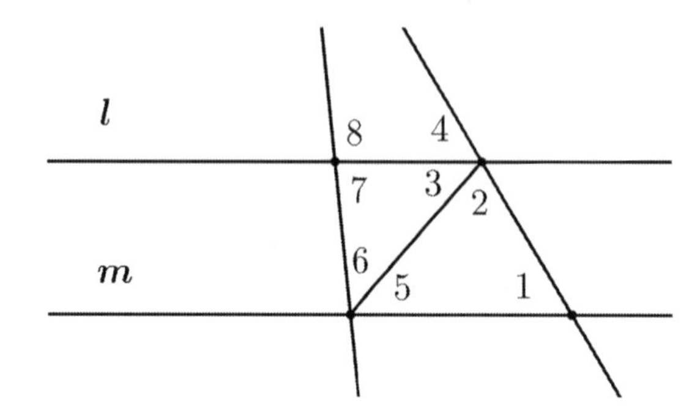

Suppose that $\angle 1 = 60^\circ, \angle 5 = 50^\circ, \angle 8 = 105^\circ$. Find the measure of $\angle 6$.

Answer

55°

Solution

Since l and m are parallel, angle 8 and the angle made up of angles 5 and 6 put together are corresponding angles. Hence

$$105^\circ = \angle 8 = \angle 6 + \angle 5 = \angle 6 + 50^\circ.$$

Therefore,

$$\angle 6 = 105^\circ - 50^\circ = 55^\circ.$$

Problem 1.7 Prove that the angles in any triangle ABC add up to 180°.

Solution

View $\overline{AB}$ as the base of the triangle and draw a line parallel to $\overline{AB}$ through C, as shown in the diagram below.

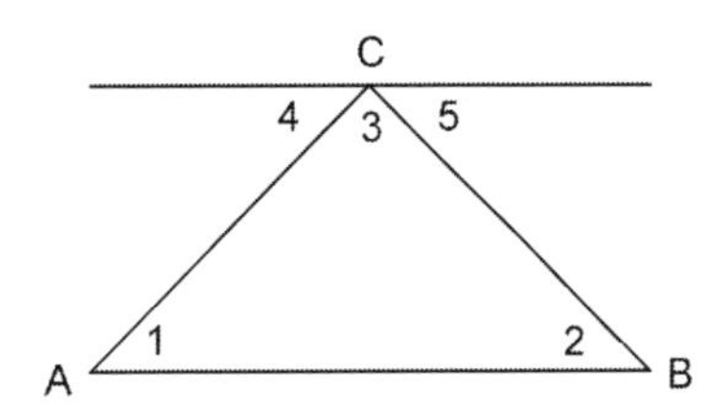

Since $\angle 3, \angle 4, \angle 5$ combine to form a straight line we have $\angle 3 + \angle 4 + \angle 5 = 180^\circ$. Then note $\angle 1 = \angle 4$ and $\angle 2 = \angle 5$ as they are alternating interior angles. Thus $\angle 1 + \angle 2 + \angle 3 = 180^\circ$ as needed.

Problem 1.8 Let $\angle A, \angle B, \angle C$ be the angles in triangle $\triangle ABC$. If $\angle B - \angle A = \angle C - \angle B$ and $\angle A = 33^\circ$, what is the angle measure of $\angle C$?

Answer

87°

Solution

Call $x = \angle B - \angle A = \angle C - \angle B$. Then the angles have measures

$$\angle A = 33, \angle B = 33 + x, \angle C = 33 + 2x.$$

The three angles must add up to 180° so

$$33 + 33 + x + 33 + 2x = 180 \text{ hence } 3x = 180 - 99 = 81$$

and $x = 27$. The largest angle will have measure $33 + 2 \times 27^\circ = 87^\circ$.

Problem 1.9 Consider the lines $y = 0$, $y = mx + 1$, and $y = \frac{-1}{m}x - 1$ where $m \neq 0$. If the lines $y = 0$ and $y = mx + 1$ form an angle of 30°, what angles are formed from the line $y = 0$ and $y = \frac{-1}{m}x - 1$?

Answer

60° and 120°

Solution

The two lines are perpendicular, so they intersect at 90° angles as shoen below.

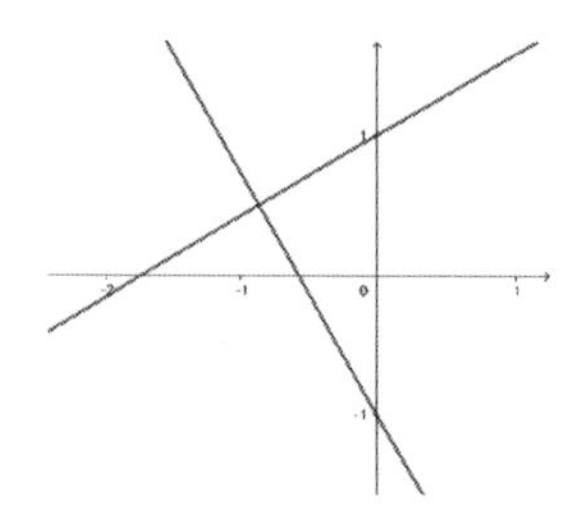

Hence they form a right triangle with the x-axis. As the triangle has angles of 30° ($y = mx + 1$ and the x-axis) and 90° ($y = mx + 1$ and $\frac{-1}{m}x - 1$) the last angle, formed by $\frac{-1}{m}x - 1$ and the x-axis, must be $180^\circ - 90^\circ - 30^\circ = 60^\circ$. The other angle formed is supplementary, which is $180^\circ - 60^\circ = 120^\circ$.

Problem 1.10 Suppose angles $\angle A$ and $\angle B$ are complementary, with ratio of angles $\angle A : \angle B = 7 : 11$. What is the ratio of the angles supplementary to $\angle A$ and supplementary to $\angle B$?

Answer

$29:25$

Solution

We know the ratio $\angle A : \angle B = 7 : 11$ so let x such that $\angle A = 7x$ and $\angle B = 11x$. As $\angle A$ and $\angle B$ are complementary we have $\angle A + \angle B = 7x + 11x = 90^\circ$ so $x = 5^\circ$. Therefore $\angle A = 35^\circ$ and $\angle B = 55^\circ$. Therefore supplementary angles are respectively $180^\circ - 35^\circ = 145^\circ$ and $180^\circ - 55^\circ = 125^\circ$ with ratio $145 : 125 = 29 : 25$.

2 Solutions to Chapter 2 Examples

Problem 2.1 A rectangle is divided into 4 smaller rectangles by two lines, as shown. The perimeters of three of these rectangles are 12, 14, and 14. Find the perimeter of the remaining (shaded) rectangle.

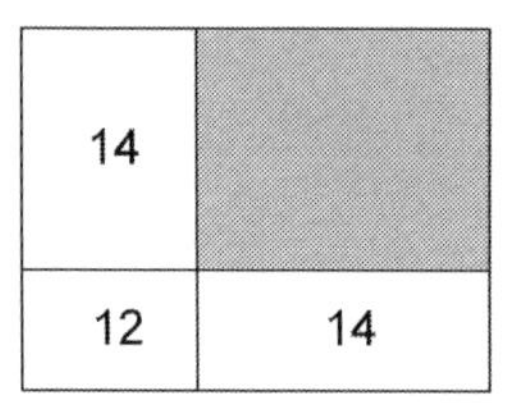

Answer

16

Solution

Let x denote the length of the lower left rectangle. If its perimeter is 12, it must have a height of $6-x$. Similarly, the upper left rectangle has dimensions x by $7-x$. As the lower right rectangle has the same height as the lower left rectangle, its dimensions must be $x+1$ by $6-x$. Therefore, the dimensions of the shaded rectangle are $x+1$ by $7-x$ hence its perimeter is 16.

Problem 2.2 A rectangle is divided into 5 squares, as shown in the diagram.

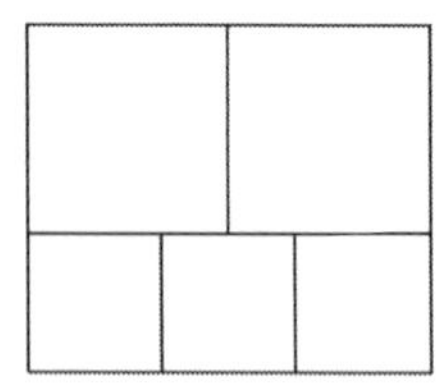

Given that the area of one bigger square is 16 in^2 more than that of one smaller square, find the area of the whole rectangle.

Answer

96

Solution

Let a be the side length of the bigger square and b be the side length of the smaller square, so $2a = 3b$ by the diagram. We are given that $a^2 = b^2 + 16$ and we want to find $2a^2 + 3b^2$. Since $2a = 3b$ we know $4a^2 = 9b^2$. Substituting $a^2 = b^2 + 30$ we have

$$4(b^2 + 16) = 9b^2 \Rightarrow 5b^2 = 64 \Rightarrow b^2 = \frac{64}{5}.$$

Therefore (again using $a^2 = b^2 + 30$),

$$2a^2 + 3b^2 = 2(b^2 + 16) + 3b^2 = 5b^2 + 32 = 64 + 32 = 96.$$

Problem 2.3 Arrange several equilateral triangles and rhombi, all of whose side lengths are 2 cm, to form a long parallelogram, as shown in the diagram.

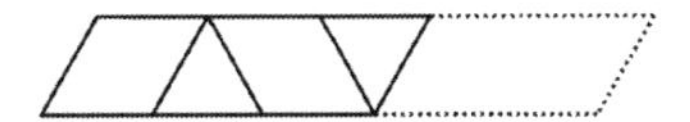

Assume the perimeter of the long parallelogram is 368 cm, how many equilateral triangle and rhombi are there?

Answer

60 triangles, 61 rhombi

Solution

The left and right sides each has length 2 cm, so the top and bottom sides each has length $\frac{368 - 2 \times 2}{2} = 182$. Each group of 2 rhombi and 2 triangles has top side with length 6 cm, so there are 30 such full groups, with 2 cm left over. Based on the pattern, this 2 cm is an extra rhombus, so there are 60 triangles and 61 rhombi.

Problem 2.4 Prove the Pythagorean Theorem using the diagram below:

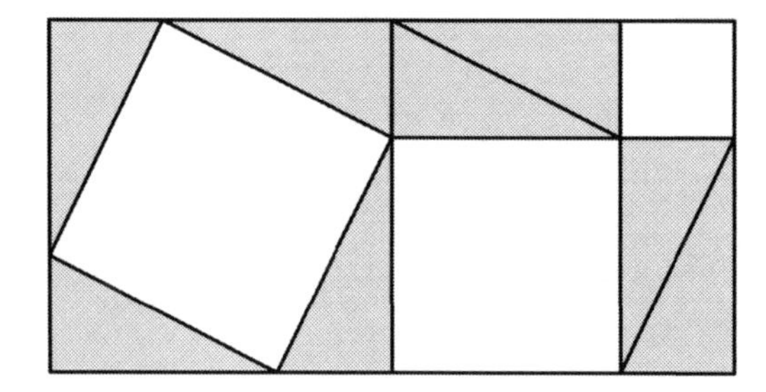

Solution

Let the triangles be denoted T, and the small, middle, and big squares denoted (respectively) S, M, B as labeled in the diagram below:

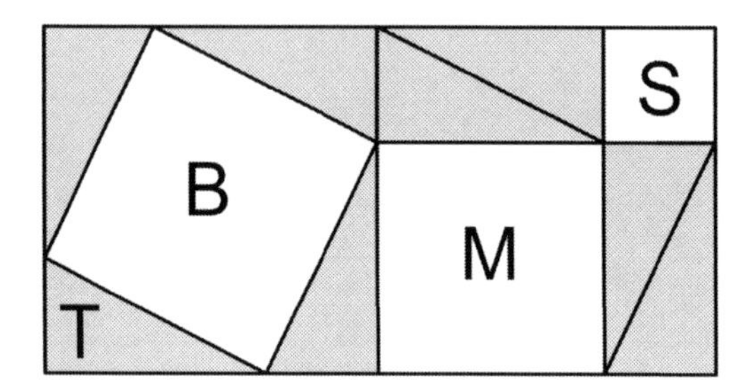

Note the entire diagram is made up of two congruent squares, so they have same area. The left square is $B+4T$ and the right square is $S+M+4T$. Hence, the area of square B is the sum of the squares M, S. If T has side lengths $a < b < c$, it is easy to see the areas of S, M, L are respectively a^2, b^2, c^2 and the result follows.

Problem 2.5 The lines $-4x+3y=2$, $-4x+3y=27$, $3x+4y=-14$, and $3x+4y=11$ intersect at the points $(-6,1)$, $(-3,5)$, $(-2,-2)$, and $(1,2)$ to form a quadrilateral.

Show that this quadrilateral is a square.

Solution

To show this quadrilateral is a square, we need to show (i) it is a rectangle, and (ii) all the sides have the same length.

For (i), note our four lines have respective slopes of $\frac{3}{4}, \frac{3}{4}, \frac{-4}{3}, \frac{-4}{3}$, so the first two are parallel and are perpendicular to the second two (the second two are parallel). Hence the four angles are all right angles as needed.

For (ii) we use the distance formula to calculate the distance of each side. For example, the distance from $(-6,1)$ to $(-3,5)$ is

$$\sqrt{(-6-(-3))^2+(1-5)^2} = \sqrt{3^2+4^2} = \sqrt{25} = 5.$$

Similarly the distances from $(-3,5)$ to $(1,2)$, from $(1,2)$ to $(-2,-2)$, and from $(-2,-2)$ to $(-6,1)$ are each 5. Therefore all four sides of the quadrilateral are equal.

Since the quadrilateral has 4 right angles and 4 equal sides it is a square.

Problem 2.6 Review of areas of parallelograms, triangles, and trapezoid. Try to understand all of these using the area of a rectangle as a starting point.

(a) The area of a parallelogram is bh.

Solution

As in the diagram below, cutting a triangle at one end of the parallelogram and moving it to the other results in a $b \times h$ rectangle.

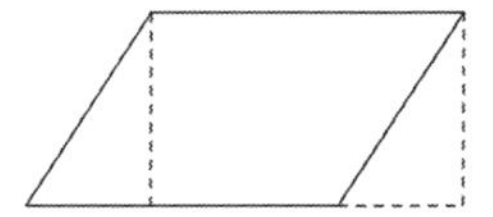

(b) The area of a triangle is $\frac{1}{2}bh$.

Solution

Method 1: Two copies of any triangle can be combined to form a parallelogram with base b and height h.

Method 2: A right triangle is half a rectangle, so has area $\frac{1}{2}bh$. Then any triangle can be split into two right triangles (by dropping an altitude to the longest side).

(c) The area of a trapezoid is $\frac{b_1+b_2}{2}h$.

Solution

Draw a diagonal of the trapezoid. This breaks the trapezoid into two triangles, each with height h and having bases b_1 and b_2.

Problem 2.7 Prove $(a+b)^2 = a^2 + 2ab + b^2$ geometrically.

Solution

Consider the diagram below:

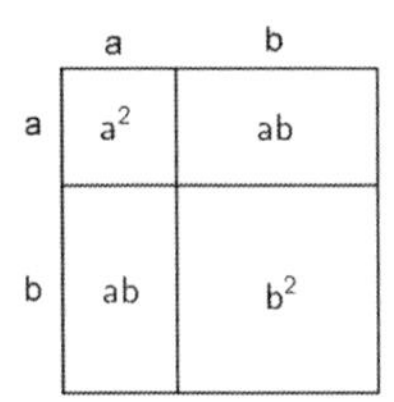

Note that the total area is $(a+b)^2$ but also $a^2 + 2ab + b^2$. Hence the two quantities must be equal.

Problem 2.8 Find the point on the line $y = 2x + 5$ that is the closest to the origin.

Answer

$(-2, 1)$

Solution

Recall the shortest distance is perpendicular to the line. Hence we find where $y = 2x+5$ intersects the line $y = -\frac{x}{2}$ (slope $\frac{-1}{2}$ containing point $(0,0)$). Solving we get $x = -2$ so $y = 1$ and the minimum distance is achieved at $(-2, 1)$.

Problem 2.9 A big rectangle is divided into 6 squares of different sizes, as shown.

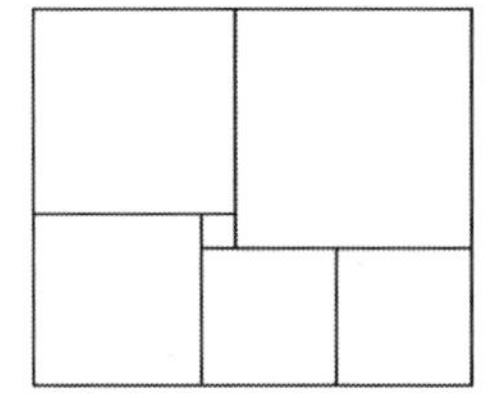

Given that the smallest square in the middle has area 1 cm^2, find the area of the big rectangle in square centimeters.

Answer

143

Solution

Let x be the side length of the square at the lower right corner. Then the side lengths of the squares (except for the smallest one), in clockwise order, are x, x, $x+1$, $x+2$, $x+3$. Comparing the top edge and the bottom edge, we have equation $x+2+x+3 = x+x+x+1$. Thus $2x+5 = 3x+1$, and solve to get $x = 4$. So the length of the rectangle is $x+x+x+1 = 13$, and the width is $(x+1)+(x+2) = 2x+3 = 11$. So the area of the big rectangle is $13 \times 11 = 143$ cm^2.

Problem 2.10 Suppose a triangle has vertices $(3,4),(4,7),(7,6)$. Find the area of the triangle.

Answer

5

Solution

Calculate the line between $(3,4),(7,6)$ as $-x+2y = 5$ with slope $\frac{1}{2}$. Hence a perpendicular line has slope -2, and the one containing $(4,7)$ is $2x+y = 15$. These two lines intersect at $(5,5)$, so using the distance formula we can calculate the altitude has length $\sqrt{5}$. As the distance between $(3,4),(7,6)$ is $2\sqrt{5}$, the area of the triangle is 5.

3 Solutions to Chapter 3 Examples

Problem 3.1 For each of the following "rules", state whether they work for proving congruence, similarity, both, or neither. If the rule does not work, give a counterexample.

(a) SAS (two sides and the angle between them)

Answer

Works for both congruence and similarity

(b) AAA (all three angles)

Answer

Works for similarity but not congruence

Solution

Note scaling a triangle (making it larger or smaller) does not change the angles.

(c) ASA (two angles and the side between them)

Answer

Works for both

Solution

Note that once we know two angles we know the third, so this could be referred to as AAA plus we know a side. We already know the triangles are similar (just from AAA), and knowing a side ensures congruence as well.

(d) AAS (two angles and a side not between them)

Answer

Works for both

Solution

Note that once we know two angles we know the third, so this could be referred to as AAA plus we know a side. We already know the triangles are similar (just from AAA), and knowing a side ensures congruence as well.

(e) SSA (two sides and an angle not between them)

Answer

Works for neither

Solution

Counterexamples may vary, for example cutting an isosceles triangle into two unequal pieces (through the third vertex).

Problem 3.2 Let triangle ABC be equilateral triangle with side length 20. Let D be on side $\overline{AB}$ and E be on side $\overline{AC}$ such that $\overline{DE} || \overline{BC}$. Assume triangle ADE and trapezoid $DECB$ have the same perimeter. What is the length of $\overline{BD}$?

Answer

5

Solution

Let $x = BD$. Note that $\triangle ADE$ is equilateral as $\overline{DE}$ is parallel to $\overline{BC}$. Thus the perimeter of $\triangle ADE = 3(20-x)$ and the perimeter of trapezoid $DECB$ is $(20-x)+2x+20 = 40+x$. Since these two perimeters are equal, $60-3x = 40+x$ so $x = 5$.

Problem 3.3 Prove the converse of the Pythagorean Theorem. Note, we have already proven the Pythagorean theorem, so we can use it in this proof!

Solution

Suppose we have triangle $\triangle ABC$ with $AB = c, AC = b, BC = a$. We want to show that if $c^2 = a^2 + b^2$ then $\triangle ABC$ is a right triangle. Now form a different right triangle $\triangle A'B'C'$ with $A'C' = b$ and $B'C' = a$. By the Pythagorean Theorem, $A'B' = c$. Hence, using SSS, $\triangle ABC \cong \triangle A'B'C'$ and therefore $\angle C'$ is right.

Problem 3.4 Let $\triangle ABC$ with $AB = 12$, $BC = 16$, $AC = 20$. Let D, E, and F be the midpoints of, respectively, AB, BC, and AC. What is the area of $\triangle DEF$?

Answer

24

Solution

First note that $(12,16,20)$ is a Pythagorean triple, so $\triangle ABC$ is right. Since D, E, and F are midpoints we know

$$AD = BD = 6, BE = CE = 8, AF = CF = 10.$$

$(6,8,10)$ is also a Pythagorean triple so as $\triangle ADF$, $\triangle DBE$, and $\triangle FEC$ are all right triangles, they each must have sides 6, 8, and 10. Therefore $DF = 8$, $EF = 6$, $DE = 10$ and $\triangle DEF$ is a right triangle with area $\frac{8 \cdot 6}{2} = 24$.

Note we actually showed the four triangles were all congruent to each other, and each similar to the full triangle with ratio of sides $1:2$.

Problem 3.5 A 'Golden Triangle' is an isosceles triangle with angles in ratio $1:2:2$.

(a) Find the angles in a golden triangle.

Answer

$36°, 72°, 72°$

Solution

The angles are in ratio $1:2:2$ so let them be $x, 2x, 2x$. Thus $x + 2x + 2x = 180°$ so $5x = 180°$ and $x = 36°$. This gives the angles are $36°$, $72°$, and $72°$.

(b) Find the ratio of the sides in a golden triangle. The diagram shown below with golden triangle ABC may help.

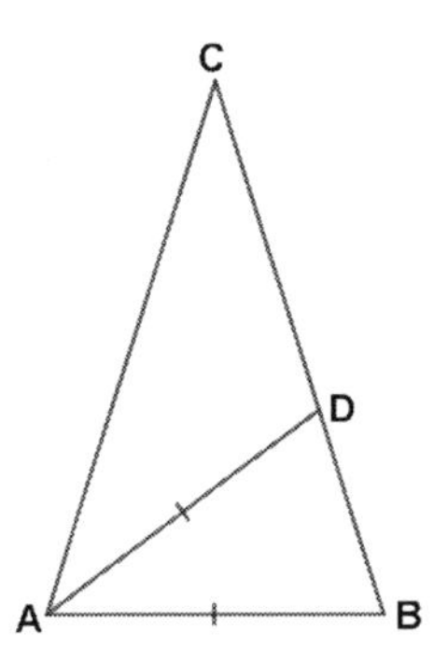

Answer

$1 : \varphi : \varphi$ where $\varphi = \dfrac{1+\sqrt{5}}{2}$ (the golden ratio)

Solution

Let $x = AB$ and $BD = 1$. Since $\triangle ABD$ is isosceles, $\angle ADB = \angle ABD = 72^\circ$, so $\angle BAD = \dfrac{180^\circ - 144^\circ}{2} = 36^\circ$. Hence $\angle CAD = 72^\circ = 36^\circ = 36^\circ$ so $\triangle DCA$ is also isosceles. Thus $CD = AD = AB = x$.

We also have that $\triangle ABD \sim \triangle CAB$ so the ratio of sides must be equal:

$$\frac{CB}{AB} = \frac{AD}{BD} \Rightarrow \frac{x+1}{x} = \frac{x}{1} \Rightarrow x^2 = x+1.$$

Solving for x we have $x^2 - x - 1 = 0$ so $x = \dfrac{1 \pm \sqrt{5}}{2}$ using the quadratic formula. $1 - \sqrt{5}$ is negative, so we must have $x = \dfrac{1 \pm \sqrt{5}}{2} = \phi$ (called the golden ratio). The ratio of sides (using $\triangle ABD$) is therefore $1 : \phi : \phi$.

Problem 3.6 What is the ratio of the sides of a triangle with angles 30°, 60°, and 90°?

Answer

$1 : \sqrt{3} : 2$

Solution

Since we are looking for a ratio, assume the smallest side of the triangle is 1. Combine two 30-60-90 triangles to form an equilateral triangle as shown below,

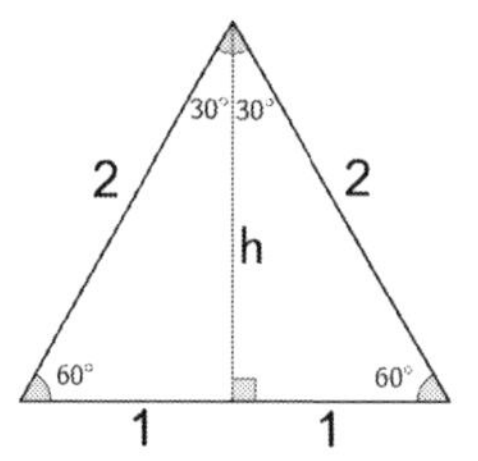

so we know the hypotenuse is $1+1=2$. If we call the missing leg h, we have $1^2+h^2=2^2$ so $h=\sqrt{3}$ using the Pythagorean theorem. As any 30-60-90 triangle is similar to this one, the ratio of sides is always $1:\sqrt{3}:2$.

Problem 3.7 Given that one of the angles of the triangle with sides $(5,7,8)$ is $60°$, show that one of the angles of the triangle with sides $(3,5,7)$ is $120°$.

Solution

Let $\triangle ABC$ be such that $AB=5, BC=7, CA=8$. Then the $60°$ angle must be $\angle A$, as it must correspond to the middle angle in the triangle. Pick the point D on BC such that $AB=AD$, then $\triangle ABD$ is equilateral, and $\triangle ADC$ is a triangle with sides $(3,5,7)$ and $\angle ADC=120°$.

Problem 3.8 Suppose A and B are two points. Describe the set of points that are equal distance from A as from B.

Answer

The points form the perpendicular bisector of $\overline{AB}$

Solution

We claim all the points on the perpendicular bisector of $\overline{AB}$ are the same distance from A and B. Clearly the midpoint M works. Let C be another point on the perpendicular bisector and consider the triangles AMC and BMC. We have $AM=BM$, $\angle AMC=90°=\angle BMC$, and $MC=MC$ so using SAS the two triangles are congruent. Hence $AC=BC$ as claimed.

Problem 3.9 $\triangle ABC$ is an equilateral triangle with with $A = (0,2)$ and $B = (\sqrt{3},1)$. Find all possible coordinates for C.

Answer

$(0,0), (\sqrt{3},3)$.

Solution

Since $\triangle ABC$ is an equilateral triangle, we must have $AC = BC = AB$ (all three sides have the same length). As C is the same distance from A and B, it must be on the perpendicular bisector of $\overline{AB}$. The midpoint of $\overline{AB}$ is $\left(\frac{\sqrt{3}}{2}, \frac{3}{2}\right)$ and the slope is $\frac{1}{-\sqrt{3}}$. Hence the perpendicular bisector has equation

$$y - \frac{3}{2} = \sqrt{3}\left(x - \frac{\sqrt{3}}{2}\right) \text{ which simplifies to } y = \sqrt{3}x.$$

Note the length $AB = \sqrt{\sqrt{3}^2 + 1^2} = 2$, so C must be distance 2 from A. We know C is of the form $(x, \sqrt{3}x)$, so using the distance formula we have

$$2 = \sqrt{(x-0)^2 + (\sqrt{3}x - 2)^2} \Rightarrow 4 = x^2 + 3x^2 - 4\sqrt{3}x + 4 \Rightarrow 4x^2 - 4\sqrt{3}x = 0.$$

Therefore $4x(x - \sqrt{3}) = 0$ so $x = 0$ or $x = \sqrt{3}$. Plugging in for the y-values we get $C = (0,0)$ or $C = (\sqrt{3},3)$.

Problem 3.10 What is the side length of the largest equilateral triangle $\triangle AEF$ that can fit inside square $ABCD$ with side length 1? For reference, $\triangle AEF$ is shown in the diagram, with $BE = DF$.

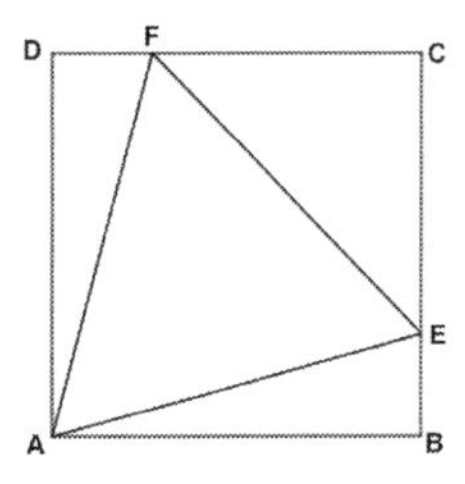

Answer

$\sqrt{6} - \sqrt{2}$.

Solution

Let $x = EC = CF$, then $BE = DF = 1 - x$. So $1^2 + (1 - x)^2 = 2x^2$, solving for x, we get $x = (\sqrt{3} - 1)$. Then as $\triangle ECF$ is a 45-45-90 triangle, the side length $EF = \sqrt{2}(\sqrt{3} - 1) = \sqrt{6} - \sqrt{2}$.

4 Solutions to Chapter 4 Examples

Problem 4.1 Similar triangles review. We use $a = BC, b = AC, C = AB$ to denote the sides of $\triangle ABC$.

(a) Let ABC be a right triangle with $\angle A = 30°$, $\angle C = 90°$, and $c = 4$. What are a and b?

Answer

$a = 2$, $b = 2\sqrt{3}$

Solution

Notice that the remaining angle is 60°, so this is a 30-60-90 triangle. Hence $a : b : c = 1 : \sqrt{3} : 2 = 2 : 2\sqrt{3} : 4$ so $a = 2$ and $b = 2\sqrt{3}$.

(b) Let ABC be a triangle with $\angle A = 15°$, $\angle C = 90°$, $a = \sqrt{3} - 1$, $b = \sqrt{3} + 1$, and $c = 2\sqrt{2}$. Let DEF be a triangle with $\angle D = 15°$, $\angle F = 90°$, and $DE = 12$. What are EF and DF?

Answer

$EF = 3\sqrt{6} - 3\sqrt{2}$ and $DF = 3\sqrt{6} + 3\sqrt{2}$

Solution

Note $\triangle ABC \sim \triangle DEF$ with ratio of sides $2\sqrt{2} : 12 = 1 : 3\sqrt{2}$. Hence $EF = 3\sqrt{2} \times (\sqrt{3} - 1) = 3\sqrt{6} - 3\sqrt{2}$ and $DF = 3\sqrt{2} \times (\sqrt{3} + 1) = 3\sqrt{6} + 3\sqrt{2}$.

Problem 4.2 The following triangles are not necessarily drawn to scale. For each of them calculate $\sin(\theta), \cos(\theta), \tan(\theta)$.

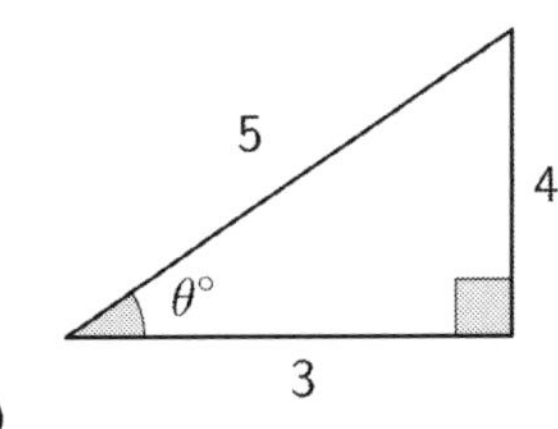

(a)

Answer

$\sin(\theta) = \frac{4}{5}, \cos(\theta) = \frac{3}{5}, \tan(\theta) = \frac{4}{3}$

(b)

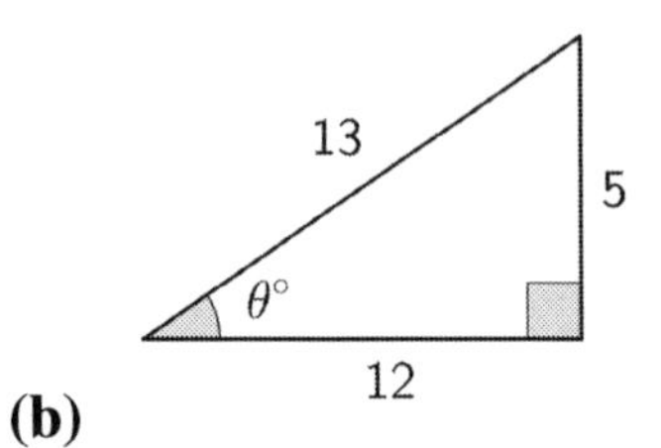

Answer

$\sin(\theta) = \frac{5}{13}, \cos(\theta) = \frac{12}{13}, \tan(\theta) = \frac{5}{12}$

Problem 4.3 Calculate

(a) $\sin(60^\circ)$

Answer

$\frac{\sqrt{3}}{2}$

Solution

This is clear from a $30-60-90$ triangle with sides 1, $\sqrt{3}$, 2.

(b) $\cos(60^\circ)$

Answer

$\frac{1}{2}$

(c) $\tan(60^\circ)$

Answer

$\sqrt{3}$

Problem 4.4 Suppose you have an angle θ in a right triangle with $\cos(\theta) = \frac{4}{5}$.

(a) Calculate $\sin(\theta)$.

Answer

$\frac{3}{5}$

Solution

Assume the opposite and hypotenuse sides of the triangle are $4, 5$. Using the Pythagorean theorem the last side must be 3.

(b) Calculate $\tan(\theta)$.

Answer

$\frac{3}{4}$

Problem 4.5 Suppose $\triangle ABC$ is a right triangle with $\angle C = 90°$. Let $BC = a$, $AC = b$, $AB = c$. Find a, b, c for each of the triangles below using the given information.

(a) $\sin(\angle A) = \frac{4}{5}$ and $c = 25$.

Answer

$a = 20, b = 15, c = 25$

Solution

$\sin(\angle A) = \frac{4}{5} = \frac{a}{c}$, so the triangle must have sides a, b, c in ratio $4, 3, 5$.

(b) $\tan(\angle A) = 1$ and $c = 4$.

Answer

$a = 2\sqrt{2}, b = 2\sqrt{2}, c = 4$

Solution

$\tan(\angle A) = 1 = \frac{a}{b}$, so the triangle must have sides a, b, c in ratio $1, 1, \sqrt{2}$.

Problem 4.6 Verify that $\tan(\theta) = \dfrac{\sin(\theta)}{\cos(\theta)}$ for $\theta = 30°$.

Solution

Consider a $30-60-90$ triangle with sides $1, 2$ and $\sqrt{3}$. Then $\sin(30°) = \frac{1}{2}$, $\cos(30°) = \frac{\sqrt{3}}{2}$, and $\tan(30°) = \frac{\sqrt{3}}{3}$. We can see then

$$\frac{\sin(30°)}{\cos(30°)} = \frac{\frac{1}{2}}{\frac{\sqrt{3}}{2}} = \frac{\sqrt{3}}{3} = \tan(30°).$$

Problem 4.7 Verify that $\sin^2(\theta)+\cos^2(\theta) = 1$ for $\theta = 60°$. Note: $\sin^2(\theta) = (\sin(\theta))^2$.

Solution

We have $\sin(60°) = \frac{\sqrt{3}}{2}$, and $\cos(60°) = \frac{1}{2}$, so

$$\sin^2(60°)+\cos^2(60°) = \left(\frac{\sqrt{3}}{2}\right)^2 + \left(\frac{1}{2}\right)^2 = \frac{3}{4}+\frac{1}{4} = 1.$$

Problem 4.8 Let ABC be a right triangle with $\angle C = 90°$. If $\sin(\angle B) = \frac{7}{25}$, and $BC = 48$ what is the length of the height from vertex C?

Answer

13.44

Solution

Since $\sin(\angle B) = \frac{7}{25}$, the sides of $\triangle ABC$ are in ratio $7 : 24 : 25$, so $AB = 50$, $BC = 48$, and $AC = 14$. Let D be the feet of the altitude from C. Then, by AAA, $\triangle ABC \sim \triangle ACD$, so $\frac{DC}{14} = \frac{48}{50}$. Therefore $DC = \frac{336}{25} = 13.44$.

Problem 4.9 Suppose you know you are 10 feet from the base of a building. You also know that the angle from the ground to the top of the building is $80°$. How tall is the building? Hint: $\tan(80°) \approx 5.67$.

Answer

≈ 56.7 feet tall

Solution

Let h be the height of the building. Then we have a right triangle formed by you, the base of the building, and the top of the building. Therefore, $\tan(80^\circ) = \frac{O}{A} = \frac{h}{10}$ so (since $\tan(80^\circ) \approx 5.67$, we have that $h \approx 10 \cdot 5.67 = 56.7$ feet.

Problem 4.10 Suppose John's cup is a cylinder with diameter 5 cm and height 12 cm. Suppose he wants to fill his cup halfway, but the water fountain he uses to fill it only shoots water to a height of 4.8 cm. Can he fill his cup halfway? Hint: Find the exact height needed to fill the cup halfway.

Answer

Yes

Solution

Note the side view of the cup is a rectangle, and for the cup to be half full we have the diagram below (where the diagonal is parallel to the ground):

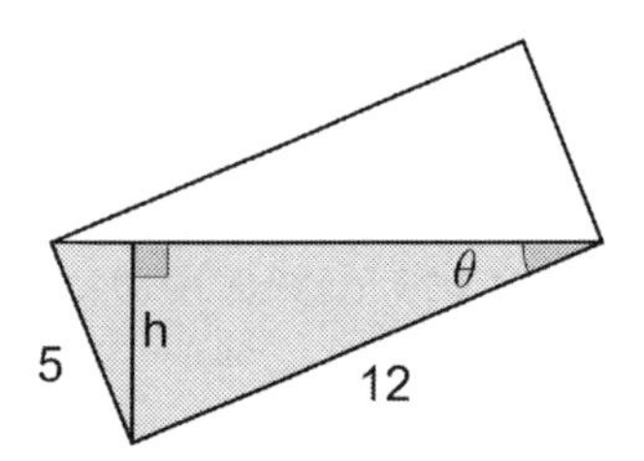

We then have that $\sin(\theta) = \frac{O}{H} = \frac{h}{12}$. The diagonal has length $\sqrt{5^2 + 12^2} = 13$, so (using the larger right triangle) $\sin(\theta) = \frac{O}{H} = \frac{5}{13}$. Hence $\frac{h}{12} = \frac{5}{13} \Rightarrow h = 5 \cdot \frac{12}{13} \approx 4.6 < 4.8$. Thus, John is able to fill his cup at least halfway.

5 Solutions to Chapter 5 Examples

Problem 5.1 We've already shown that the interior angles of a triangle add up to 180°. Explain a general formula for the sum of the interior angles of a polygon with n sides.

Answer

$180^\circ(n-2) = 180n^\circ - 360^\circ$

Solution

The angles of a triangle add up to 180°. A polygon with n sides can be divided into $n-2$ triangles, where the sum of the angles in all the triangles combined equals the sum of the angles in the polygon. Hence the angles sum up to $180^\circ(n-2)$ as needed.

Problem 5.2 Exterior Angles

(a) Prove that the exterior angles of a triangle add up to 360°.

Solution

If the interior angles of a triangle are x, y, z then the exterior angles are $180^\circ - x, 180^\circ - y, 180^\circ - z$. Therefore the sum of the exterior angles is

$$180^\circ - x + 180^\circ - y + 180^\circ - z = 540^\circ - (x+y+z).$$

As the interior angles add up to 180°, $x+y+z = 180^\circ$ and thus the sum of the exterior angles is $540^\circ - 180^\circ = 360^\circ$.

(b) Explain a general formula for the sum of the exterior angles in a polygon with n sides.

Answer

360°

Solution

The method from part (a) can be expanded to work for polygons of n sides.

Alternatively, pretend the polygon's sides are streets, and you are walking along the streets. Each time you turn a corner, you turn an exterior angle. As you walk around all the streets, the total of your turns is the sum of all the exterior angles. However, when you return to the starting point, you are facing the same direction that you started in, except you have turned around once. Since one full turn is 360°, the sum of the exterior angles is 360° as well.

Problem 5.3 Complete the following table about polygons with n sides: name, sum of interior angles, sum of exterior angles, and measure of each angle in case of regular polygon. All angles are in degrees. Justify your answers. Keep the chart for your own reference.

n	Name	Int. Angle Sum	Ext. Angle Sum	Each Angle (if regular)
3	Triangle			
4				
5				
6				
7	Heptagon			
8				
9	Nonagon			
10				
12	Dodecagon			
20	Icosagon			

Solution

Using the previous two problems we can fill in the table as below:

n	Name	Int. Angle Sum	Ext. Angle Sum	Each Angle (if regular)
3	Triangle	$180°$	$360°$	$60°$
4	Quadrilateral	$360°$	$360°$	$90°$
5	Pentagon	$540°$	$360°$	$108°$
6	Hexagon	$720°$	$360°$	$120°$
7	Heptagon	$900°$	$360°$	$900/7°$
8	Octagon	$1080°$	$360°$	$135°$
9	Nonagon	$1260°$	$360°$	$140°$
10	Decagon	$1440°$	$360°$	$144°$
12	Dodecagon	$1800°$	$360°$	$150°$
20	Icosagon	$3240°$	$360°$	$162°$

Problem 5.4 Four non-overlapping regular plane polygons all have sides of length 1. The polygons meet at a point A in such a way that the sum of the four interior angles at A is $360°$. Among the four polygons, two are squares and one is a triangle. What is the perimeter of the entire shape?

Answer

9.

Solution

First note that the fourth polygon must have an interior angle of $360° - 90° - 90° - 60° = 120°$ and hence must be a hexagon. Therefore the perimeter of the entire shape is the sum of the perimeter of each of the four polygons minus the 4 shared edges. Note each shared edge is counted twice. Hence the perimeter is $6+4+4+3-4\cdot 2=9$.

Problem 5.5 Consider the quadrilateral $ABCD$ shown below.

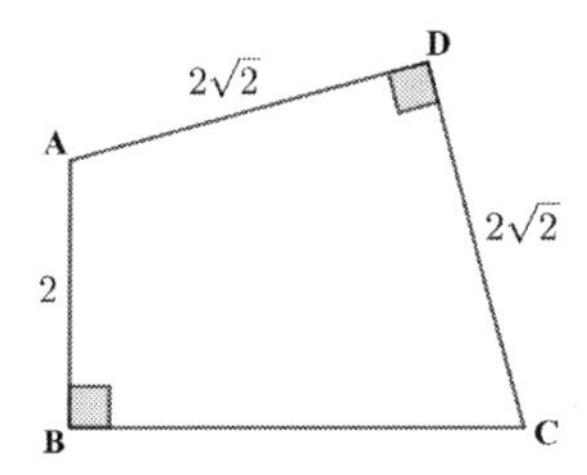

Find the missing side BC as well as the measures of $\angle A$ and $\angle C$.

Answer

$BC = 2\sqrt{3}, \angle A = 105°, \angle C = 75°$

Solution

Connect A to C. Then $\triangle ADC$ is an isosceles right triangle, so is a 45-45-90 triangle. Thus $AC = 2\sqrt{2}\cdot\sqrt{2} = 4$. Looking then at $\triangle ABC$, it is a right triangle with side 2 and hypotenuse 4, so the remaining side $BC = 2\sqrt{3}$. Therefore $\triangle ABC$ is a 30-60-90 triangle. From this it follows that $\angle A = 60° + 45° = 105°$ and $\angle C = 30° + 45° = 75°$.

Problem 5.6 Area of Triangles for SAS

(a) Find the area of $\triangle ABC$ if $BC = 5$, $AC = 6$, and $\angle C = 60°$.

Answer

$$\frac{15\sqrt{3}}{2}$$

Solution

Consider AC the base and draw an altitude as in the diagram below.

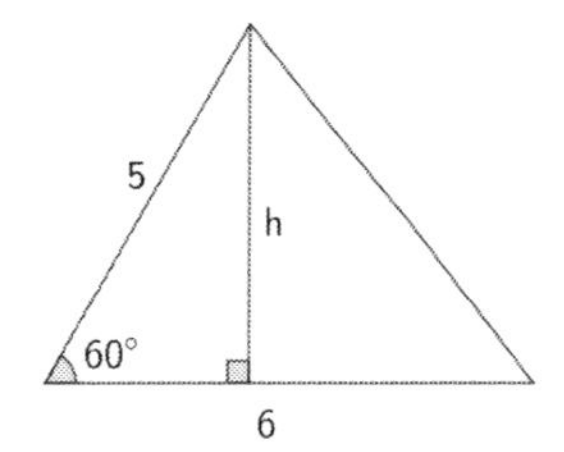

We have that

$$\sin(60^\circ) = \frac{h}{5} \text{ implying that } h = 5\sin(60^\circ) = 5\cdot\frac{\sqrt{3}}{2} = \frac{5\sqrt{3}}{2}.$$

Thus the area of the triangle is $\frac{1}{2}\cdot 6\cdot\frac{5\sqrt{3}}{2} = \frac{15\sqrt{3}}{2}$.

(b) Extend your method in part a) to give a formula for the area of $\triangle ABC$ if you know $BC = a$ and $AC = b$ and $\angle C = \theta$.

Answer

$\frac{1}{2}ab\sin(\theta)$

Solution

If we consider a the base, we can draw an altitude as in part (a). In general we have $\sin(\theta) = \frac{h}{b}$ so $h = b\sin(\theta)$. Thus the area of $\triangle ABC$ is $\frac{1}{2}ab\sin(\theta)$.

Problem 5.7 A square is formed by putting 4 congruent isosceles right triangles in the corners. The shaded square is the region not covered by the triangles as shown in the diagram below.

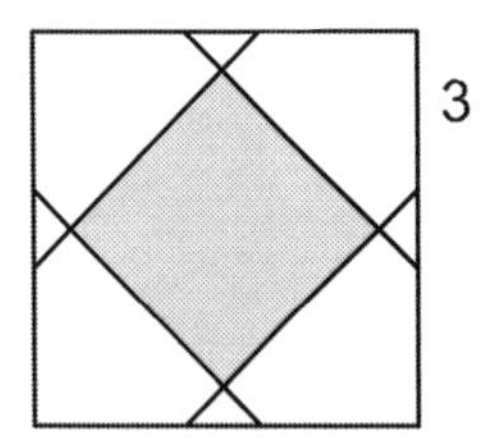

Caution: 3 is not the side length of the isosceles triangle. What is the area of the shaded square?

Answer

18

Solution

Note the opposite hypotenuses of the triangles are parallel. Hence, if we create an isosceles triangle with side length 3, we see its hypotenuse has the same length as the square. Hence, the square has side length $3\sqrt{2}$ and area 18.

Problem 5.8 In parallelogram $ABCD$ as shown, $BC = 10$. Triangle BCE is a right triangle where $\overline{BE}$ is the hypotenuse, and $EC = 8$. Given that $[ABG]+[CDF]-[EFG] = 10$, find the length of $\overline{CF}$.

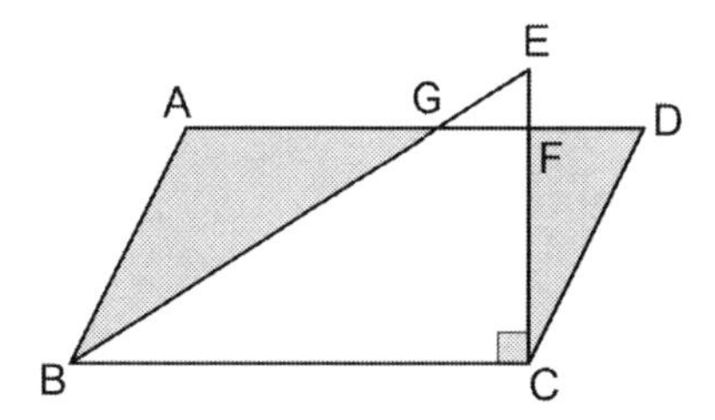

Answer

5

Solution

$[ABCD]-[EBC] = [ABG]+[CDF]+[BCFG]-[EFG]-[BCFG] = 10$, thus $BC \cdot CF - \frac{1}{2}BC \cdot CE = 10$. So $10 \cdot CF - \frac{1}{2}10 \cdot 8 = 10$, solve and get $CF = 5$.

Problem 5.9 In equiangular octagon $ABCDEFGH$, $AB = CD = EF = GH = 6\sqrt{2}$ and $BC = DE = FG = HA$. Given the area of the octagon is 184, compute the length of side BC.

Answer

4.

Solution 1

Let $x = BC$, connect $\overline{AD}, \overline{EH}, \overline{BG}, \overline{CF}$. The sides $\overline{AB}, \overline{CD}, \overline{EF}, \overline{GH}$ are hypotenuses of four isosceles right triangles, whose legs are equal to 6. Thus the area of the whole octagon is

$$x^2 + 4 \cdot 6 \cdot x + 4 \cdot \frac{1}{2} \cdot 6 \cdot 6 = 184,$$

so

$$x^2 + 24x - 112 = 0.$$

We want the positive root: $x = 4$.

Solution 2

A better solution involves extending the sides $\overline{BC}, \overline{DE}, \overline{FG}, \overline{HA}$ to both directions to make a big square. The added corners of this square are isosceles right triangles with hypotenuses $6\sqrt{2}$, so their legs are all 6. Let $x = BC$, the big square's side length is $x + 12$. The added corners have a total area of two squares of side 6, so the added area is 72. Thus $(x+12)^2 = 184 + 72 = 256$, thus $x + 12 = 16$, and $x = 4$.

Problem 5.10 Inscribed Regular Polygons

(a) Find the area of the largest equilateral triangle that fits inside a circle of radius 1. Note, this triangle is said to be inscribed in a circle of radius 1.

Answer

$$\frac{3\sqrt{3}}{4}$$

Solution

Using the center of the circle, we can divide the equilateral triangle into 3 congruent circles as shown in the diagram below.

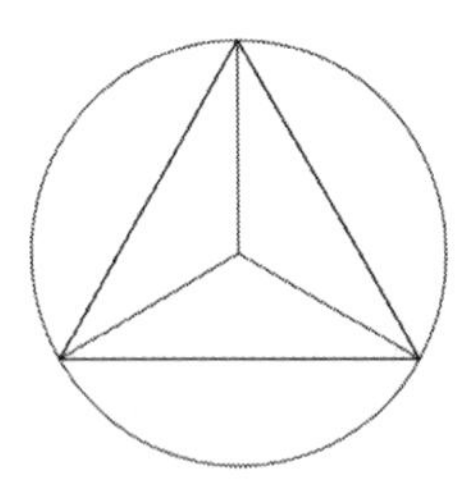

Note that the central angle of each triangle is $360° \div 3 = 120°$. Hence each of the 3 triangles has two sides of length 1 with an $120°$ angle in between. We can use our formula from earlier in class to find the area of each of these triangles as

$$\frac{1}{2} \cdot 1 \cdot 1 \cdot \sin(120°) = \frac{\sqrt{3}}{4}.$$

(Alternatively, each triangle can be broken into two 30-60-90 triangles.) Hence the total area is $3 \times \frac{\sqrt{3}}{4} = \frac{3\sqrt{3}}{4}$.

(b) Extend your method from part (a) to give a general formula for the area of a regular polygon with n sides inscribed inside a circle of radius 1.

Answer

$\frac{n}{2}\sin\left(\frac{360^\circ}{n}\right)$

Solution

Similar to part (a), we can divide the polygon into n congruent triangles, each with two sides of length 1 and an angle of $\frac{360^\circ}{n}$ in between. Hence the area is

$$n \cdot \left(\frac{1}{2} \cdot 1 \cdot 1 \cdot \sin\left(\frac{360^\circ}{n}\right)\right) = \frac{n}{2}\sin\left(\frac{360^\circ}{n}\right).$$

6 Solutions to Chapter 6 Examples

Problem 6.1 Consider the circle given by the equation $x^2 - 6x + y^2 + 4y = 0$.

(a) Find the center and radius of the circle by completing the square.

Answer

$(3, -2)$ with radius $\sqrt{13}$

Solution

Completing the square we have $(x-3)^2 + (y+2)^2 = 13$ so the center is $(3, -2)$ and the radius is $\sqrt{13}$.

(b) What is the area and circumference of the circle?

Answer

Area: 13π, Circumference: $2\pi\sqrt{13}$

Solution

We know the radius is $\sqrt{13}$ from part (a) so the area is $\pi(\sqrt{13})^2 = 13\pi$ and the circumference is $2\pi\sqrt{13}$.

Problem 6.2 Suppose you have a circle $(x-1)^2 + (y-1)^2 = 4$ and a line $x + y = 4$.

(a) The line and circle intersect at two points A, B, find them.

Answer

$(1,3), (3,1)$.

Solution

Of course you can just solve the system of equations. One way to guess and check is as follows. The two numbers must sum to 4, and it is easy to see that $(3,1), (1,3)$ both work $(2^2 + 0^2 = 4)$. Because we know geometrically that a circle and line can intersect in at most two points, these must be all the solutions.

(b) Verify that the perpendicular bisector of $\overline{AB}$ goes through the center of the circle.

Solution

You can calculate the perpendicular bisector of $\overline{AB}$ to be $y = x$, and clearly the center of the circle $(1,1)$ is on it.

Problem 6.3 Let $\angle APB$ be an inscribed angle on a circle with center O. Prove that $\angle APB$ is half the angular size of arc $\widehat{AB}$ if:

(a) O lies on $\angle APB$.

Solution

Assume O lies on $\overline{AP}$. Then $\triangle OPB$ is isosceles, so $2\angle BPO = 2\angle BPA = \angle BOA$ as needed.

(b) O lies inside $\angle APB$.

Solution

Let Q be such that $\overline{PQ}$ is a diameter. Note $\triangle BOP$ is isosceles, so $\angle BOQ = 2\angle BPO$. Similarly, $\angle AOQ = 2\angle APO$. Hence, $\angle BPA = \angle BPO + \angle APO = (\angle BOQ + \angle AOQ)/2 = \angle BOA/2$ as needed.

Problem 6.4 Prove that if two chords AC, BD intersect outside a circle at point P then the measure of $\angle APB$ is half the difference of the angular sizes of $\widehat{AB}, \widehat{CD}$.

Solution

Let the points be as in the diagram earlier (so C, D are on $\overline{AP}, \overline{BP}$). Connect $\overline{BC}$. Then using $\triangle ACB$ we have $\angle APB + \angle CBD = \angle ACB$. Therefore, $\angle APB = \angle ACB - \angle CBD$. Since $\angle ACB, \angle CBD$ are inscribed angles with respective arcs $\widehat{AB}, \widehat{CD}$, the result follows.

Problem 6.5 Suppose $\widehat{AB}$ and $\widehat{CD}$ are arcs each with angular size 50° and if rays $\overrightarrow{BA}, \overrightarrow{DC}$ are extended to intersect at a point E (so A is on $\overline{BE}$ and on $\overline{DE}$), $\angle AEC = 50^\circ$. Find the angular size of arc $\widehat{BD}$. Express your answer in degrees, rounded to the nearest tenth if necessary.

Answer

180

Solution

Let the angular measure of $\widehat{BD} = x$. Then the size of $\widehat{AC} = 360° - 50° - 50° - x = 260° - x$. Then $\angle AEC = 50° = \frac{x - (260° - x)}{2}$ so solving for x gives $x = 180°$.

Problem 6.6 Points A, B, C, and D are on a circle forming quadrilateral $ABCD$. If $\angle A = 2x$, $\angle B = 3x$, $\angle C = 5y$, and $\angle D = 3y$, what are x and y?

Answer

$x = 40°, y = 20°$

Solution

Consider first $\angle A$ and $\angle C$. We have $\angle A = \angle BAD$ and $\angle C = \angle BCD$. Therefore

$$\angle A = \frac{1}{2} m\widehat{BCD} \text{ and } \angle C = \frac{1}{2} m\widehat{BAD}$$

where, for example, $\widehat{BCD}$ is the arc containing C. Hence

$$\angle A + \angle C = \frac{1}{2} m\widehat{BCD} + \frac{1}{2} m\widehat{BAD} = \frac{1}{2}(m\widehat{BCD} + m\widehat{BAD}) = \frac{1}{2} \cdot 360° = 180°.$$

An identical argument gives $\angle B + \angle D = 180°$. Thus we have $2x + 5y = 180°$ and $3x + 3y = 180°$. From the second equation we know $2x + 2y = 120°$ so subtracting this from the first equation we have $3y = 60°$ so $y = 20°$. Finally, $3x + 3 \cdot 20° = 180°$ so $x = 40°$.

Problem 6.7 Points B and C are on a circle. A is such that $\overline{AB}$ and $\overline{AC}$ are both tangent to the circle. Prove that $AB = AC$.

Solution

Let O be the center of the circle and draw radii $\overline{OB}$ and $\overline{OC}$. Connecting $\overline{OA}$, $\triangle ABO$ and $\triangle ACO$ are right triangles as $\overline{AB}$ and $\overline{AC}$ are tangent to the circle. As $\triangle ABO$ and $\triangle ACO$ share side $\overline{OA}$ and $OB = OC$ (both are radii) we have $\triangle ABO \cong \triangle ACO$ using Hypotenuse-Leg. Therefore $AB = AC$ as needed.

Problem 6.8 A circle of radius AB is tangent to $\overline{BC}$ which has length $2\sqrt{5} + 2$. $\overline{AC}$ intersects the circle at point D so that $CD = 4$ as shown in the diagram below.

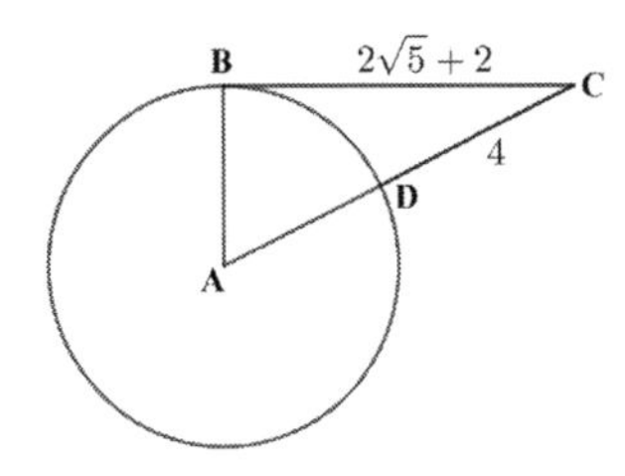

What is the area of the circle?

Answer

$(2\sqrt{5}+6)\pi$

Solution

$\overline{AB}$ and $\overline{AD}$ are both radii, so let $AB = AD = r$. $\overline{BC}$ is tangent, so $\triangle ABC$ is a right triangle. Thus

$$AB^2 + BC^2 = AC^2 \text{ meaning } r^2 + (2\sqrt{5}+2)^2 = (r+4)^2.$$

Expanding we have $r^2 + 20 + 8\sqrt{5} + 4 = r^2 + 8r + 16$ so after simplifying we have $8\sqrt{5} + 8 = 8r$. Hence $r = \sqrt{5} + 1$. Therefore the area of the circle is $\pi(\sqrt{5}+1)^2 = \pi(5 + 2\sqrt{5} + 1) = (2\sqrt{5}+6)\pi$.

Problem 6.9 Consider the parallel lines $y = x + 4$ and $y = x - 4$. What is the equation of the largest circle with center $(0,0)$ that is tangent to both lines?

Answer

$x^2 + y^2 = 8$

Solution

The lines $y = x + 4$ and $y = x - 4$ are parallel. As the circle has center $(0,0)$, we have a graph as in the diagram shown below.

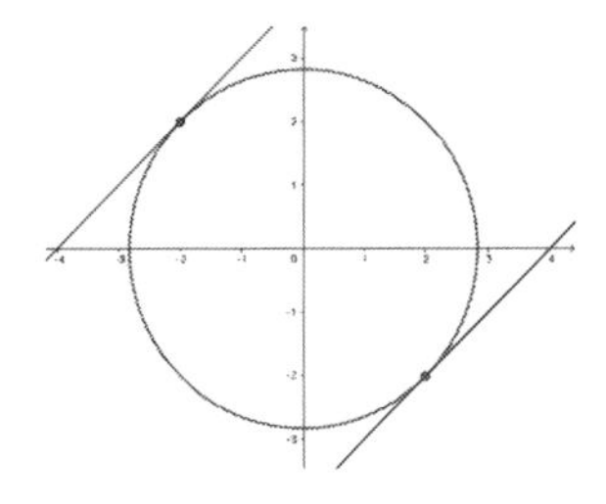

As any tangent line is perpendicular to a radius of the circle, connecting the two points of tangency gives the diameter shown below.

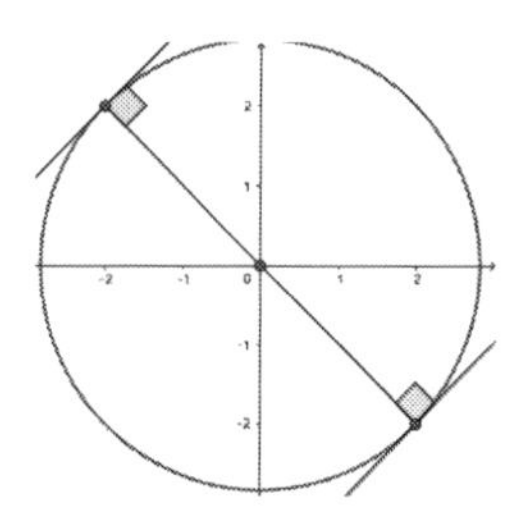

As this diameter is perpendicular to $y = x + 4$ and $y = x - 4$ (both with slope 1) it must be a part of the line $y = -x$. Substituting we can solve for when this line intersects the tangent lines

$$-x = x + 4 \text{ and } -x = x - 4 \text{ so they intersect when } x = \pm 2.$$

Hence the points of intersection are $(-2, 2)$ and $(2, -2)$ so the radius is

$$\sqrt{(2-0)^2 + (-2-0)^2} = \sqrt{8}.$$

Therefore the circle has equation $x^2 + y^2 = 8$.

Problem 6.10 Find all k such that $y = kx$ is tangent to the circle $(x-4)^2 + (y-2)^2 = 4$.

Answer

$0, \dfrac{4}{3}$

Solution

The circle has center $(4, 2)$ with radius $\sqrt{4} = 2$ so the circle is tangent to the x-axis $y = 0$. Hence $k = 0$ is one such k. There is one more as shown in the diagram below.

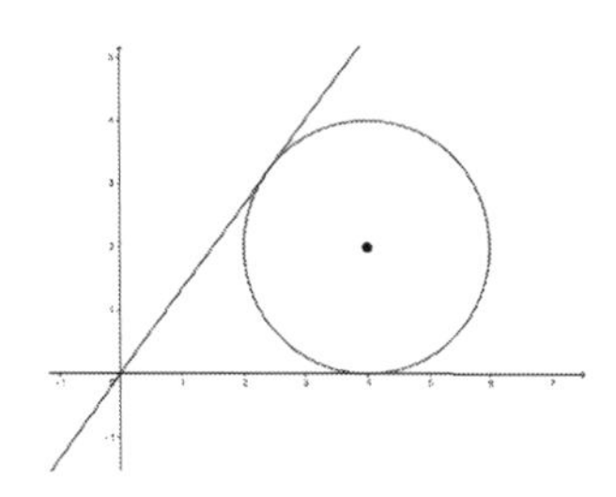

$y = 0$ is tangent to the circle at $(4,0)$, so line segment connecting the origin to the tangent point has length 4. As $y = 0$ and $y = kx$ always intersect at $(0,0)$, any tangent line will intersect the circle at distance 4 from the origin. Hence if (x,y) is an intersection point, we know

$$x^2 + y^2 = 16 \text{ and } (x-4)^2 + (y-2)^2 = 4.$$

Expanding the equation of the circle we have $x^2 - 4x + 16 + y^2 - 4y + 4 = 4$ so substituting $x^2 + y^2 = 16$ we have

$$16 - 8x + 16 - 4y + 4 = 4 \text{ implying } -8x - 4y = -32 \text{ or } y = -2x + 8.$$

Substituting we have $x^2 + (-2x+8)^2 = 16$ so $5x^2 - 32x + 64 = 16$. Solving this we have $x = 4$ or $x = \frac{12}{5}$. Solving for y we get intersection points of $(4,0)$ or $\left(\frac{12}{5}, \frac{16}{5}\right)$.
Hence

$$k = 0 \text{ or } k = \frac{16/5}{12/5} = \frac{4}{3}.$$

7 Solutions to Chapter 7 Examples

Problem 7.1 In two-dimensional geometry (inside a plane), two different lines are either parallel or they intersect.

(a) Come up with a definition of 'parallel' that makes sense in three-dimensions.

Solution

Descriptions may vary, but the general idea is that both lines should point in the same direction. Equivalently, the lines do not intersect and it is possible to form a plane containing both lines.

(b) Argue that two different lines in space can be non-intersecting and not parallel at the same time. (Such lines are called *skew* lines.)

Solution

Answers may vary. One way to produce such an example is to start with intersecting lines in a plane, and then lift one of the lines up in space so they no longer intersect.

Problem 7.2 The two equations $x+y+z=1$ and $x-y=0$ each give an equation of a plane.

(a) Graph the intersection of these planes when $x=0$, when $y=0$, and when $z=0$.

Solution

Label the equations (i) $x+y+z=1$ and (ii) $(x-y=0)$.

When $x=0$: (i) becomes $y+z=1$ and (ii) becomes $y=0$.

When $y=0$: (i) becomes $x+z=1$ and (ii) becomes $x=0$.

When $z=0$: (i) becomes $x+y=1$ and (ii) becomes $x-y=0$.

These are graphed below, with (i) dotted line and (ii) with the dashed line:

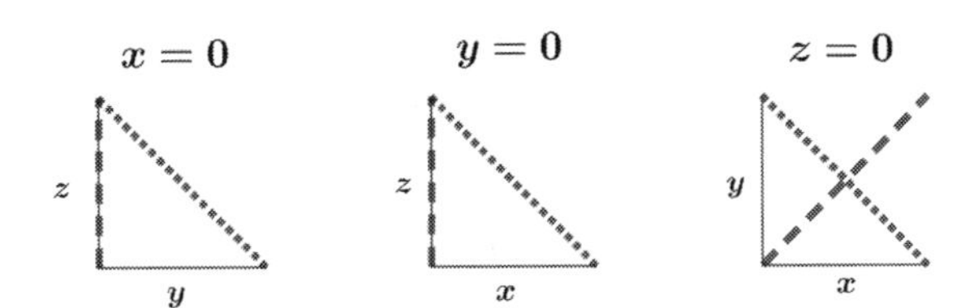

(b) The two planes intersect to form a line. Show that the expression $x = y = \dfrac{z-1}{-2}$ determines this line.

Solution

The second equation tells us that $x = y$. Substituting this into the first equation we have

$$x + x + z = 1 \text{ so solving we have } x = \frac{1-z}{2} = \frac{z-1}{-2},$$

as needed.

Problem 7.3 Distance Formula in Three-dimensions: Distance from $(0,0,0)$ to (x,y,z).

(a) If $A = (0,0,0)$, $B = (x,y,z)$, and $C = (x,y,0)$ argue that $\triangle ACB$ is a right triangle.

Solution

The line segment $\overline{AC}$ is in the xy-plane and B is directly above (meaning only the z-component changes) C. Therefore $\overline{BC}$ is perpendicular to $\overline{AC}$ and $\triangle ACB$ is a right triangle.

(b) By first calculating AC and BC, use the Pythagorean theorem to show that $AB = \sqrt{x^2 + y^2 + z^2}$.

Solution

We have $AB^2 = AC^2 + BC^2$ using the Pythagorean theorem. $BC = z$ and using the Pythagorean theorem again we have $AC^2 = x^2 + y^2$. Therefore $AB^2 = AC^2 + BC^2 = x^2 + y^2 + z^2$ and hence $AB = \sqrt{x^2 + y^2 + z^2}$.

Problem 7.4 You have a cube with side length 1. Label the vertices of the square $ABCD - EFGH$ where $ABCD$ forms the "bottom" square and $EFGH$ forms the "upper" square as in the diagram below.

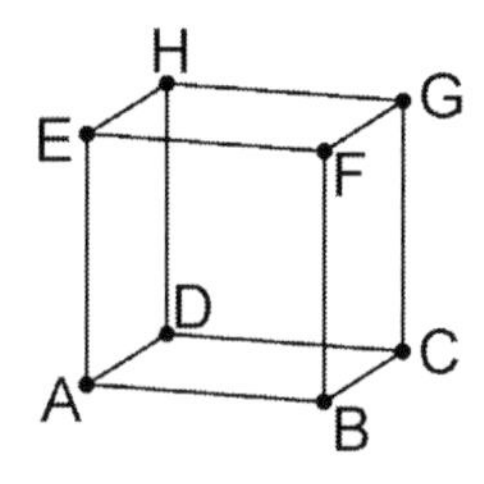

Find the distance from A to B, from A to F, and from A to G.

Answer

$AB = 1, AF = \sqrt{2}, AG = \sqrt{3}$

Solution

Consider the cube $ABCD - EFGH$ in the 3-dimensional coordinate plane with $A = (0,0,0)$, $B = (1,0,0)$, $D = (0,1,0)$, and $E = (0,0,1)$. We then have $F = (1,1,0)$ and $G = (1,1,1)$ so $AB = 1$, $AF = \sqrt{1^2 + 1^2} = \sqrt{2}$, and $AG = \sqrt{1^2 + 1^2 + 1^2} = \sqrt{3}$ using the distance formula.

Problem 7.5 How many vertices, edges, and faces does a cube have? Describe the faces.

Answer

Vertices: 8, Edges: 12, Faces: 6

Solution

Each of the faces is a square. There are 6 faces. Counting we see there are 8 vertices and 12 edges.

Problem 7.6 Euler's Theorem

(a) Euler's Theorem states that $F - E + V$ is a constant for all polyhedron. Here F is the number of faces, V the number of vertices, and E the number of edges. Using cubes as an example, what is this constant?

Answer

2

Solution

A cube has 6 faces, 12 edges, and 8 vertices, so $F-E+V=6-12+8=2$.

(b) Verify Euler's Theorem for a regular dodecahedron which is formed by 12 regular pentagons and shown below.

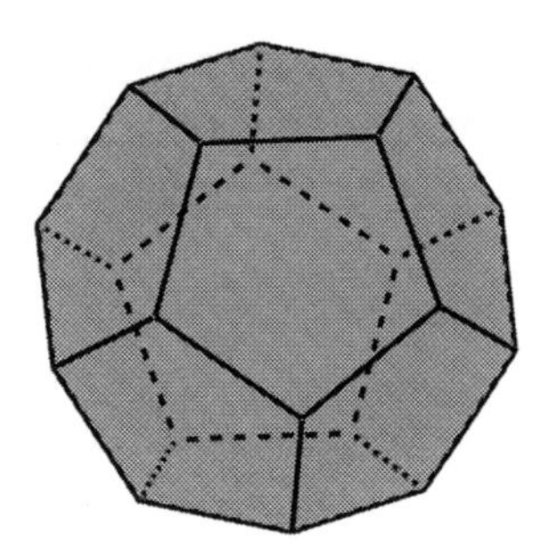

Solution

We know there are 12 faces (the pentagons). Each pentagon has 5 edges, but each of these edges is shared with a second pentagon, so there are $12\times 5\div 2=30$ edges. Similarly each pentagon has 5 vertices, but each vertex is shared with two other pentagons, so there are $12\times 5\div 3=20$ vertices.

Verifying Euler's Theorem we have $F-E+V=12-30+20=2$ which matches the theorem.

Problem 7.7 You have a box (rectangular prism) that is 2 feet long, 1 foot wide, and has a height of 6 inches.

(a) How much space is inside the box? That is, what is the volume of the box?

Answer

1 ft^3

Solution

In feet, the box has dimensions $2\times 1\times .5$ so it has volume 1 ft^3 (cubic feet).

(b) What is the surface area of the box?

Answer

7 ft^2

Solution

The surface area is $2(2\times1+2\times0.5+1\times.5)=7\ \text{ft}^2$ (square feet).

(c) You want to double the volume of the box by changing one of the dimensions of the box. What are the possible new surface areas (measured in square feet)?

Answer

13, 12, or 10

Solution

Since we are interested in determining the dimensions of the box after its volume was doubled after changing the length of one side, we observe that the possible dimensions of the new figure are $4\times1\times.5$ or $2\times2\times.5$ or $2\times1\times1$ if we are restricted to only changing one of the sides of the prism.

Therefore, the surface areas can be computed as follows:

$$2(4\times1+4\times.5+1\times.5)=13$$
$$2(2\times2+2\times.5+2\times.5)=12$$
$$2(2\times1+2\times1+1\times1)=10$$

These are the three possible surface areas of the rectangular prism after its area is doubled.

Problem 7.8 A cube is increased to form a new cube so that the surface area of the new cube is 4 times that of the original cube. By what factor is the side length increased? What about the volume of the cube?

Answer

Side Length: 2, Volume: 8

Solution

A cube consists of 6 square faces. If the surface area of the new cube is 4 times that of the original, the area of each of the square faces should increase by 4. Therefore the side length of each square should double. As the volume of a cube is s^3 where s is the side length, the volume will increase by a factor of $2^3 = 8$.

Problem 7.9 Consider a ball (or sphere) with radius 6.

(a) Find the volume of the ball using the formula $\frac{4}{3}\pi r^3$.

Answer

288π

Solution

The volume is $\frac{4\pi}{3}6^3 = 288\pi$.

(b) Find the surface area of the ball using the formula $4\pi r^2$.

Answer

144π

Solution

The surface area is $4\pi 6^2 = 144\pi$.

Problem 7.10 If a solid ball with radius 6 is cut in half, find the volume and surface area of the half-ball.

Answer

Volume: 114π, Surface Area: 108π

Solution

The volume is simply half the volume of the full ball, so $\frac{4\pi}{3}6^3 \div 2 = 228\pi \div 2 = 114\pi$.

For the surface area, we have half of the original surface area, $4\pi 6^2 \div 2 = 144\pi \div 2 = 72\pi$. However, we also have an extra circle of radius 6 giving an extra $\pi 6^2 = 36\pi$ of surface. Hence the total surface area of the half-ball is $72\pi + 36\pi = 108\pi$.

8 Solutions to Chapter 8 Examples

Problem 8.1 Cutting a rectangular prism in half along the diagonal will form a triangular prism as shown below.

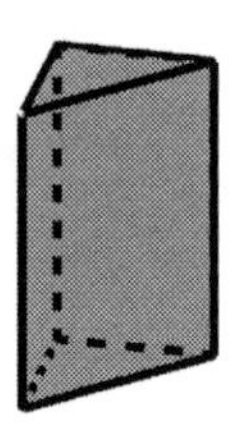

The triangular prism shown has a height of 2 cm and the base is an isosceles right triangle with legs 1 cm long.

(a) Find the volume of the prism. Explain how to extend your method to a general formula.

Answer

1 cm^3

Solution

The volume of the triangular prism is half that of a 1 by 1 by 2 rectangular prism. Hence the volume is $\frac{1}{2} \cdot 1 \cdot 1 \cdot 2 = 1$.

For a more general method, the base has area $\frac{1}{2} \cdot 1 \cdot 1 = \frac{1}{2}$ and the height is 2. Multiplying we get a volume of $\frac{1}{2} \cdot 2 = 1$.

(b) Find the surface area of the prism. Does your method extend to a general formula?

Answer

$5+2\sqrt{2}$

Solution

We can calculate the surface area by adding up the areas of each face. The top and bottom are each triangles with area $\frac{1}{2} \cdot 1 \cdot 1 = \frac{1}{2}$. The three sides are all rectangles with heights of 2. Two of these rectangles have lengths of 1, while the last has length $\sqrt{1^2+1^2} = \sqrt{2}$ (the hypotenuse of the triangle). Adding everything up we get

$$2 \cdot \frac{1}{2} + 2 \cdot (1 \cdot 2) + \sqrt{2} \cdot 2 = 5 + 2\sqrt{2}$$

as the surface area of the full prism.

Problem 8.2 Consider the cylinder (a circular prism) shown below with height 3 and radius 1.

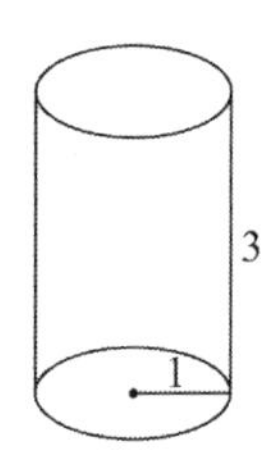

(a) What is the volume of the cylinder?

Answer

3π

Solution

The base is a circle with radius 1, so has area $\pi \cdot 1^2 = \pi$. The height is 3, so the volume is $\pi \cdot 3 = 3\pi$.

(b) What is the surface area of the cylinder?

Answer

8π

Solution

The bases are circles with radii 1, so they have a combined area of $2 \cdot \pi \cdot 1^2 = 2\pi$. Unrolling the side of the cylinder gives a rectangle that has width 3 (the height of the cylinder) and length $2\pi \cdot 1 = 2\pi$ (the circumference of the circular base). Therefore the total surface area of the cylinder is $2\pi + 3 \cdot 2\pi = 8\pi$.

Problem 8.3 Consider the shapes of a pyramid or a cone. They fit inside a rectangular prism or a cylinder as shown below:

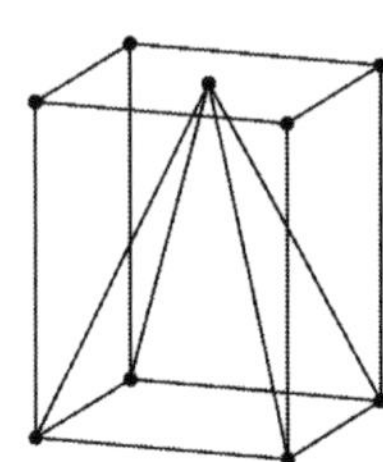
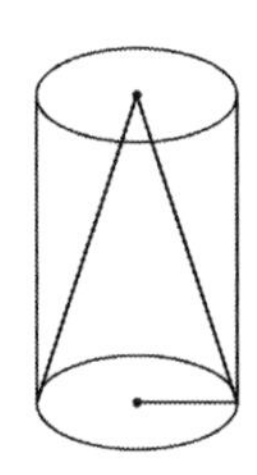

In general, the volumes of pyramids and cones are $\frac{1}{3}$ the volume of the full shape.

(a) Find the volume of a square right pyramid with base of side length 4 and height 5.

Answer

$\frac{80}{3}$

Solution

The base is a square with side length 4, so it has area $4^2 = 16$. As the height is 5, the volume is $\frac{1}{3} \cdot 16 \cdot 5 = \frac{80}{3}$.

(b) Find the volume of a cone with radius 3 and height 4.

Answer

12π

Solution

The base is a circle with radius 3, so it has area $\pi \cdot 3^2 = 9\pi$. As the height is 4, the volume is $\frac{1}{3} \cdot 9\pi \cdot 4 = 12\pi$.

(c) Find the volume of a triangular pyramid with height 3 and an equilateral triangle base with side length 3.

Answer

$\dfrac{9\sqrt{3}}{4}$

Solution

The base is an equilateral triangle with side length 3, so it has area $\dfrac{3^2\sqrt{3}}{4} = \dfrac{9\sqrt{3}}{4}$. As the height is 3, the volume is $\dfrac{1}{3} \cdot \dfrac{9\sqrt{3}}{4} \cdot 3 = \dfrac{9\sqrt{3}}{4}$.

Problem 8.4 Find the surface area of a square right pyramid with base of side length 4 and height 5.

Answer

$16 + 8\sqrt{29}$

Solution

note the square base has area

$$4^2 = 16.$$

The four triangular faces of the pyramid are congruent, each with a base of 4, but we need to calculate the height of these triangles. (Note this height is different than the height of the pyramid, and is referred to as the slant height of the pyramid.) Consider the diagram below with the slant height L.

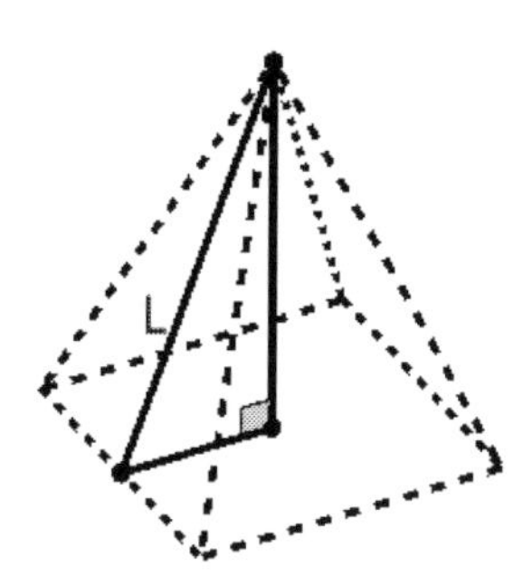

Note that the midpoint of a side, the center of the square, and the apex form a right triangle whose hypotenuse is the slant height. The other sides of this triangle are 2 (half the side length of the square) and 5 (height of the pyramid), so

$$L^2 = 2^2 + 5^2$$

so $L = \sqrt{29}$. Hence each triangle has base 4 and height $\sqrt{29}$ so each triangle has area $2\sqrt{29}$, for a total surface area of $4 \times 2\sqrt{29}$ for the triangular faces. Hence the total surface area is $16 + 8\sqrt{29}$ adding the square base.

Problem 8.5 Find the surface area of a cone with radius 3 and height 4.

Answer

24π

Solution

The bottom of the cone is a circle with area $\pi \cdot 3^2 = 9\pi$.

For the surface area of the side of the cone, consider the diagram below, where L is called the lateral height.

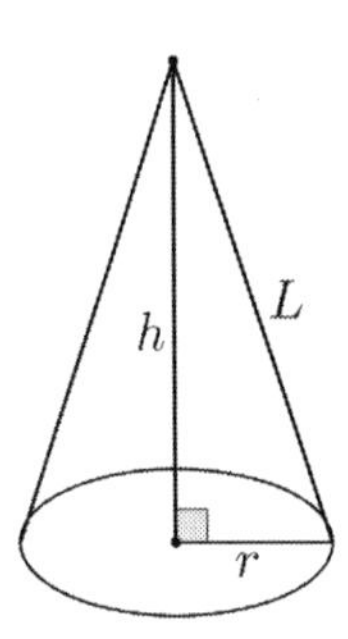

In our case $r = 3$ and $h = 4$ so $L = 5$ using Pythagorean triples. Note if we cut the side of the cone (along the lateral height) and flatten out the surface area, we get the sector shown below with radius L

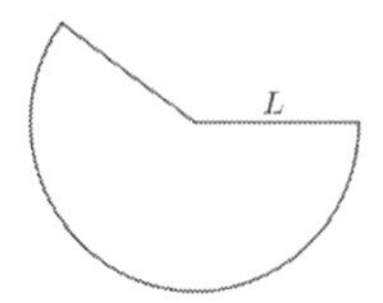

Further, the arc length of the sector must equal the circumference of the circular base of the cone. Hence this sector must be $\dfrac{2\pi \cdot 3}{2\pi \cdot 5} = \dfrac{3}{5}$ of the entire circle. This gives an area of $\dfrac{3}{5} \cdot \pi \cdot 5^2 = 15\pi$.

Hence the cone has a total surface area of $9\pi + 15\pi = 24\pi$.

Problem 8.6 Suppose you have an ice cream cone with radius 2 inches and height 4 inches. The cone starts full of ice cream (but there is not ice cream outside the cone). After you've eaten some ice cream and some of the cone you are left with a cone with a radius and a height of 2 inches. What fraction of the ice cream have you eaten?

Answer

$\frac{7}{8}$

Solution

First note that using the volume formula for the cone, there is

$$\frac{\pi}{3} \times 2^2 \times 4 = \frac{16\pi}{3}$$

cubic inches of ice cream before any is eaten. After you've eaten some ice cream, consider the following side view (where ABC is the original cone, and ADE is the "half cone").

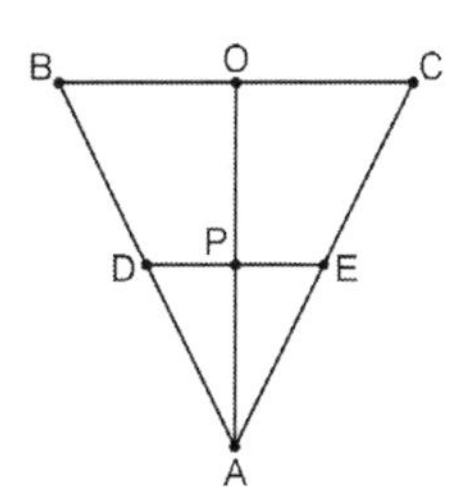

Note that $\triangle AOC \sim \triangle APE$ as they are both right triangles that share angle $\angle OAC$. Since $OA = 4, PA = 2$, we have that the ratio of sides is 2, so

$$PE = \frac{1}{2}OC = 1.$$

Hence the new cone has radius 1, and thus volume

$$\frac{\pi}{3} \times 1^2 \times 2 = \frac{2\pi}{3}$$

cubic inches of ice cream. Hence you are left with $1/8$th of the ice cream you started with, so you have eaten $7/8$th of the ice cream.

Problem 8.7 Inscribing Spheres and Cubes

(a) Find the volume of the largest sphere that fits in a cube of volume 1. (That is, inscribe a sphere inside the cube.)

Answer

$\frac{1}{6}\pi.$

Solution

The sphere will touch all 6 faces of the cube (in the middle of each face). Therefore, the diameter of the sphere must be 1. This gives a volume of $\frac{4}{3}\pi\left(\frac{1}{2}\right)^3 = \frac{1}{6}\pi.$

(b) Find the volume of the smallest sphere that holds a cube of volume 1. (That is, circumscribe a sphere outside the cube.)

Answer

$\frac{\sqrt{3}}{2}\pi.$

Solution

All of the vertices of the cube will touch the sphere. Hence the diameter of the must be $\sqrt{3}$, the length of the diagonal of the cube. his gives a volume of $\frac{4}{3}\pi\left(\frac{\sqrt{3}}{2}\right)^3 = \frac{\sqrt{3}}{2}\pi.$

Problem 8.8 Consider a unit cube $ABCD-EFGH$. Cut the cube through $\triangle AFH$ into two smaller solids.

Describe these two solids. How many faces, edges, and vertices does each have? What is the volume of each?

Answer

Triangular pyramid with volume $\frac{1}{6}$ and remaining solid has volume $\frac{5}{6}$

Solution

The smaller solid cut off from the cube is triangular pyramid $A-EFH$, with 4 faces (all triangles), 6 edges, and 4 vertices. It has height 1 with base area $\frac{1}{2}\cdot 1\cdot 1=\frac{1}{2}$. Hence its volume is $\frac{1}{3}\cdot\frac{1}{2}\cdot 1=\frac{1}{6}$.

The other solid therefore has volume $1-\frac{1}{6}=\frac{5}{6}$. This solid has 7 faces (3 squares and 4 triangles), 12 edges, and 7 vertices.

Problem 8.9 Consider a unit cube $ABCD-EFGH$. Cut the cube through $\triangle AFH$ into two smaller solids.

What is the surface area of the remaining two solids?

Answer

$\frac{\sqrt{3}}{2}+\frac{3}{2}$ and $\frac{\sqrt{3}}{2}+\frac{9}{2}$

Solution

One solid is a triangular pyramid and the other is the remaining part of the cube. Note that $\overline{AF}$, $\overline{AH}$, and $\overline{FH}$ are all face diagonals and thus $AF=AH=FH=\sqrt{2}$. Thus the area of $\triangle AFH$ is $\frac{\sqrt{2}^2\sqrt{3}}{4}=\frac{\sqrt{3}}{2}$.

For the triangular pyramid, the surface area consists of $\triangle AFH$ with area $\frac{\sqrt{3}}{2}$ and 3 congruent triangles that all have area $\frac{1\cdot 1}{2}=\frac{1}{2}$. Hence the surface area is

$$\frac{\sqrt{3}}{2}+3\cdot\frac{1}{2}=\frac{\sqrt{3}}{2}+\frac{3}{2}.$$

For the other solid, the surface area consists of $\triangle AFH$ with area $\frac{\sqrt{3}}{2}$, 3 congruent triangles that all have area $\frac{1\cdot 1}{2}=\frac{1}{2}$, and 3 squares each with area 1. Therefore its surface area is

$$\frac{\sqrt{3}}{2}+3\cdot\frac{1}{2}+3=\frac{\sqrt{3}}{2}+\frac{9}{2}.$$

Problem 8.10 An obtuse triangle with dimensions 9, 10, and 17 is rotated about the smallest side so that it creates a three-dimensional solid shown below.

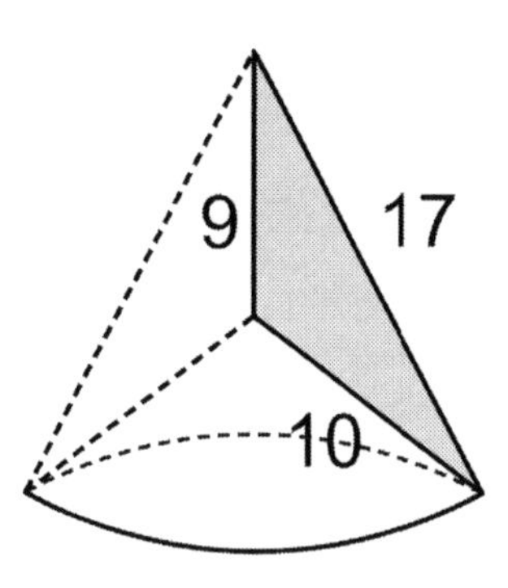

Determine the volume of the solid.

Answer

192π

Solution

Note that in the figure above, there are two cones sharing the same circular base. Let r be the radius of the cones and let h be the height of the smaller cone. Therefore, h and r satisfies

$$h^2 + r^2 = 10^2$$

and

$$(h+9)^2 + r^2 = 17^2.$$

If you recall Pythagorean triples, $h = 6$ and $r = 8$ yields a $6-8-10$ and $8-15-17$ Pythagorean triples. Therefore, the volume of the new figure is

$$\frac{1}{3}\pi \times 8^2 \times 15 - \frac{1}{3}\pi \times 8^2 \times 6 = 192\pi.$$

9 Solutions to Chapter 9 Examples

Problem 9.1 Consider the graph of the equation $x^2+y^2=z^2$ in three dimensions.

(a) What does the graph look like if $x=0$? What about if $y=0$?

Solution

If $x=0$ we have $y^2=z^2$ or, solving for z, $z=\pm y$. Similarly if $y=0$ we have $z=\pm x$. For example, the graph when $x=0$ looks like

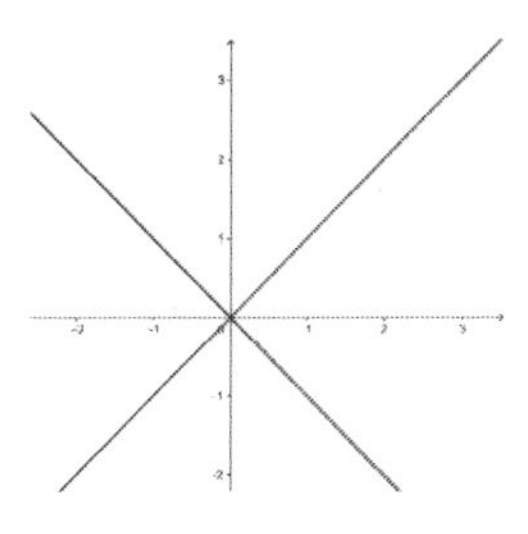

when graphed in the yz-plane.

(b) Describe the graph if $z=c$ for a constant c.

Solution

If $z=c$ for a constant c, we have $x^2+y^2=c^2$, which is a circle with radius c centered at $(0,0)$. For example, if $c=2$ the graph looks like

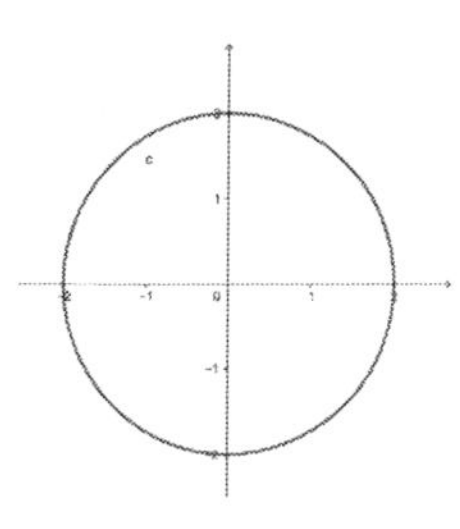

when graphed in the xy-plane.

(c) Based on the graphs from parts (a) and (b) (which are often called cross-sections), what geometric object is graphed by the equation?

Answer

Cone or Double Cone

Solution

The side views (for example when $x = 0$ or $y = 0$ are in the shape of an X. The slices when z is a constant are all circles. This gives a shape that is a double cone, as seen in the graph below.

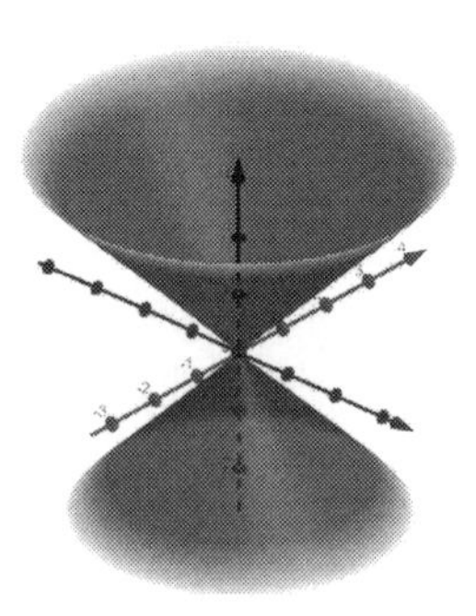

Problem 9.2 Find the volume of the solid region formed by the inequality $x^2 + y^2 \geq 4z^2$ when $0 \leq z \leq 9$.

Answer

972π

Solution

The shape of $x^2 + y^2 = 4z^2$ is a cone (as if z is a constant we get circular cross-sections). The tip of the cone is when $z = 0$ at $(0,0,0)$. When $z = 9$ we have $x^2 + y^2 = 4 \cdot 9^2 = 18^2$ so is a circle with radius 18.

Therefore we want to find the volume of a cone with radius 18 and height 9, which is $\frac{1}{3} \cdot \pi \cdot 18^2 \cdot 9 = 972\pi$.

Problem 9.3 Consider the double cone $x^2 + y^2 = z^2$.

We saw that cross sections when z was a constant were circles. Find equations and describe the graph for the cross sections when x is constant or when y is constant.

Solution

If $x = c$ we have $c^2 + y^2 = z^2$ or $z^2 - y^2 = c^2$. (Similarly we get $z^2 - x^2 = c^2$ if $y = c$.) This shape is called a hyperbola. For example, in the yz-plane the graph when $x = 2$ is given by

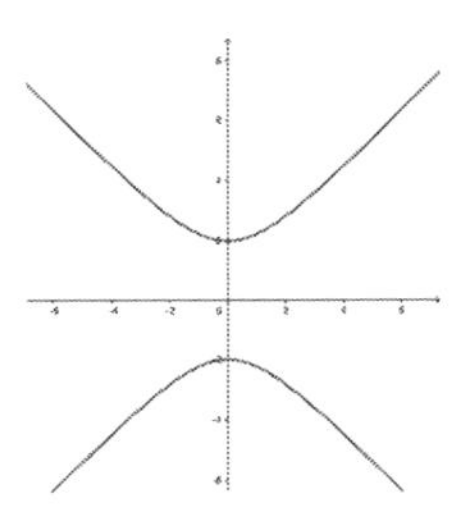

Problem 9.4 Consider the double cone $x^2 + y^2 = z^2$.

Find the equation and describe the graph of the cross section we get when $z = y + 1$.

Answer

$y = \dfrac{x^2}{2} - \dfrac{1}{2}$

Solution

Substituting $z = y + 1$ into the equation, we have $x^2 + y^2 = (y+1)^2$ so expanding we have $x^2 + y^2 = y^2 + 2y + 1$. Note the y^2 terms cancel leaving $2y = x^2 - 1$ or $y = \dfrac{x^2}{2} - \dfrac{1}{2}$. We recognize this as a parabola in the xy-plane, whose graph is shown below

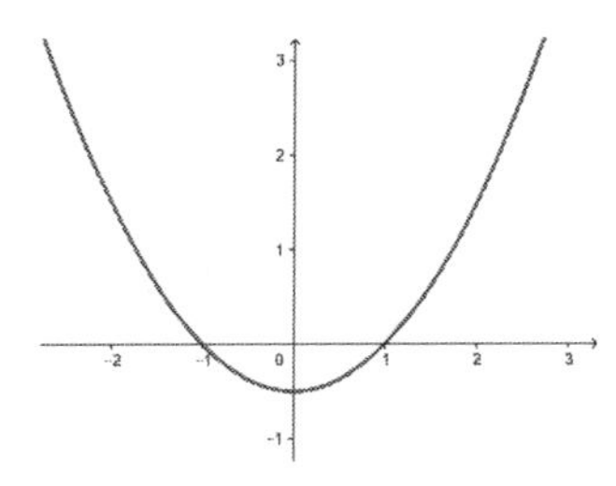

Problem 9.5 Consider the double cone $x^2 + y^2 = z^2$.

Find the equation and describe the graph of the cross section we get when $z = \dfrac{y}{2} + 1$.

Answer

$\frac{3}{4}x^2 + \frac{9}{16}\left(y - \frac{2}{3}\right)^2 = 1$

Solution

Substituting $z = \frac{y}{2} + 1$ into the equation we have

$$x^2 + y^2 = \left(\frac{y}{2} + 1\right)^2 \text{ or } x^2 + y^2 = \frac{y^2}{4} + y + 1 \text{ or } 4x^2 + 4y^2 = y^2 + 4y + 4$$

after expanding and clearing denominators. Isolating the y terms and completing the square we have

$$3\left(y^2 + \frac{4}{3}y\right) = 3\left(y^2 + \frac{4}{3}y + \frac{4}{9}\right) - \frac{4}{3} = 3\left(y - \frac{2}{3}\right)^2 - \frac{4}{3}.$$

Therefore our equation is

$$4x^2 + 3\left(y - \frac{2}{3}\right)^2 = \frac{16}{3} \Rightarrow \frac{3}{4}x^2 + \frac{9}{16}\left(y - \frac{2}{3}\right)^2 = 1.$$

This is called an ellipse, whose graph in the xy-plane is shown below.

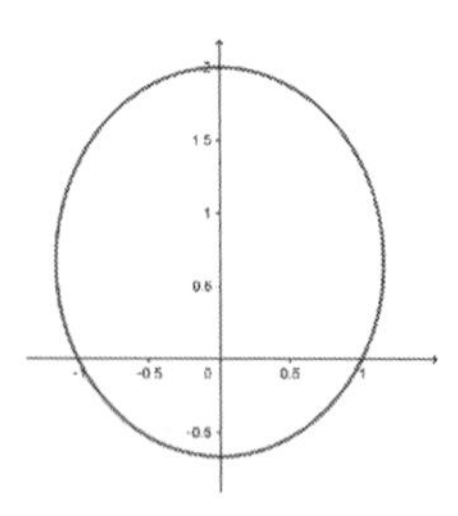

Note that the ellipse is centered at $\left(0, \frac{2}{3}\right)$, not at $(0,0)$.

Problem 9.6 Consider the double cone $x^2 + y^2 = z^2$ and the plane $z = my + 1$. Describe the values of m which produce a circle, ellipse, parabola, and hyperbola.

Answer

Circle: $m = 0$, Ellipse: $|m| < 1$, Parabola: $|m| = 1$, Hyperbola: $|m| > 1$

Solution

Algebraically we have

$$x^2+y^2=(my+1)^2=m^2y^2+2my+1 \Rightarrow x^2+(1-m^2)y^2-2my=1.$$

If $m=0$ we get $x^2+y^2=1$ which is a circle. If $|m|<1$ then $1-m^2>0$ so the coefficient of y^2 is positive, and therefore we have an ellipse. If $|m|>1$ then $1-m^2<0$ so the coefficient of y^2 is negative, and therefore we have a hyperbola. Lastly, if $m=1$, them $1-m^2=0$ so we have $x^2 \pm 2y=1$, which is a parabola.

Graphically, we look at the yz-plane pictured below, where the cone becomes $y^2=z^2$ or $z=\pm y$ and the plane $z=my+1$ is a line through $(y,z)=(0,1)$.

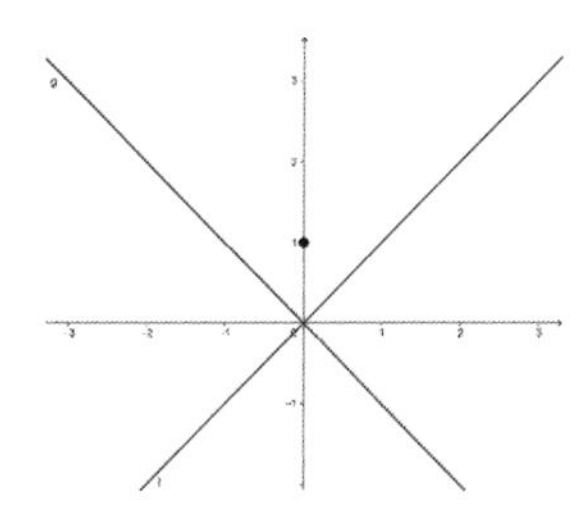

Note if $|m|<1$ then the line intersects the cone only when $z>0$, giving us an ellipse (or cone when $m=0$). When $|m|=1$ the line is parallel to the sides of the cone, giving us a parabola. If $|m|>1$ the line intersects the cone when z is both positive and negative, producing a hyperbola.

Problem 9.7 Graph the ellipse with equation $\dfrac{x^2}{4}+\dfrac{y^2}{9}=1$.

Solution

Setting $x=0$ we have $\dfrac{y^2}{9}=1$ so $y=\pm 3$. Similarly if $y=0$ we have $\dfrac{x^2}{4}=1$ so $x=\pm 2$. Plotting these points and knowing that our graph is an ellipse (an oval) we have:

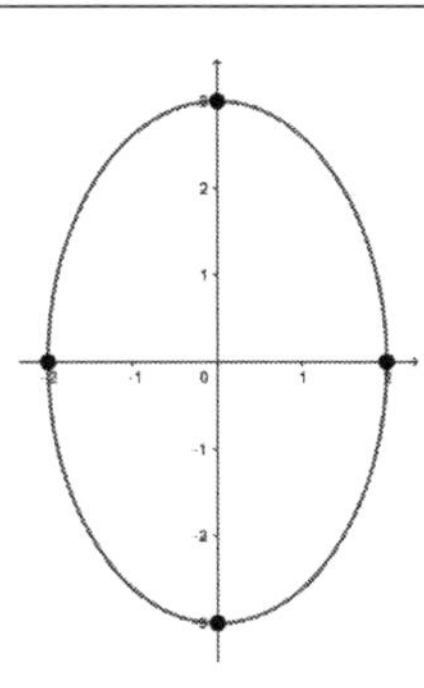

Problem 9.8 Consider the hyperbola with equation $\frac{x^2}{4} - \frac{y^2}{9} = 1.$

(a) Argue that when x and y are large the equation of the hyperbola can be approximated by $y = \pm\frac{3}{2}x$.

Solution

If we solve the equation for x we have

$$\frac{x^2}{4} = \frac{y^2}{9} + 1 \Rightarrow x = \pm 2\sqrt{\frac{y^2}{9} + 1}.$$

If y is large, the $+1$ inside the square root does not change the value much, so we have $x \approx \pm\frac{2y}{3}$. For example, if y is ± 300 the actual value of x is ± 200.01 while the approximation is ± 200.

Therefore we have $x \approx \pm\frac{2y}{3}$ or $y \approx \pm\frac{3x}{2}$ for large x and y.

(b) Use the approximation from part (a) to graph the hyperbola.

Solution

Using the approximation from part (a) we know the graph will approach $y = \pm\frac{3x}{2}$ for large x and y. Plotting a few points, if $y = 0$ we have $\frac{x^2}{4} = 1$ so $x = \pm 2$. If $y = 3$ we have

$$\frac{x^2}{4} - \frac{3^2}{9} = 1 \Rightarrow \frac{x^2}{4} = 2 \Rightarrow x = \pm\sqrt{8} = 2\sqrt{2}.$$

Therefore the hyperbola opens left/right with a graph that is shown below.

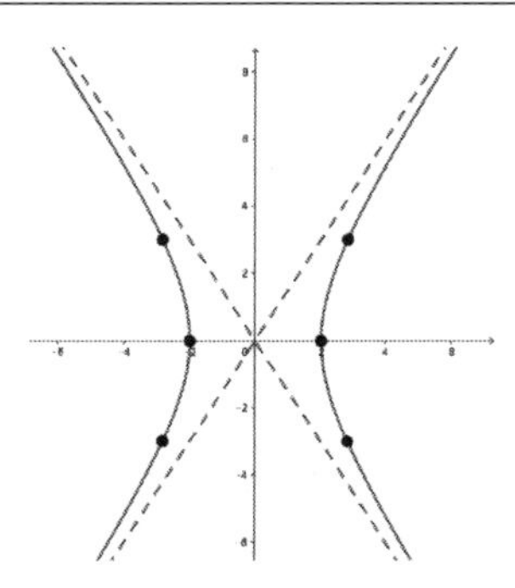

Problem 9.9 The graph of the equation $x^2 + y^2 = z$ is referred to as a paraboloid. What are the cross sections when x, y, or z are constant? Describe the graph.

Solution

If x or y is a constant we have $z = x^2 + M$ or $z = y^2 + N$ for numbers M or N. Regardless of the constant for x or y this is a parabola.

If $z = K$ is a constant, we have $x^2 + y^2 = K$ which is an circle.

This gives the graph shown below

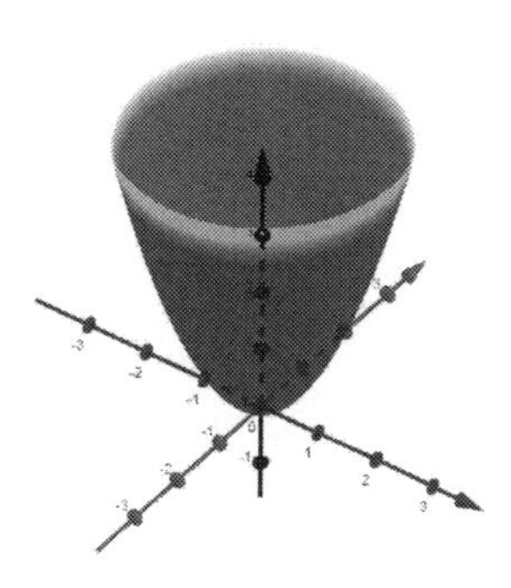

Problem 9.10 The graph of the equation $x^2 + y^2 = z^2 + 1$ is referred to as a one-sheeted hyperboloid. What are the cross sections when x, y, or z are constant? Describe the graph.

Solution

If x or y is a constant we have $z^2 - x^2 = M$ or $z^2 - y^2 = N$ for a numbers M or N. Regardless of the constant for x or y this is a hyperbola.

If $z = K$ is a constant, we have $x^2 + y^2 = K^2 + 1$. Note this is a circle for any value of K (including 0).

This gives the graph shown below

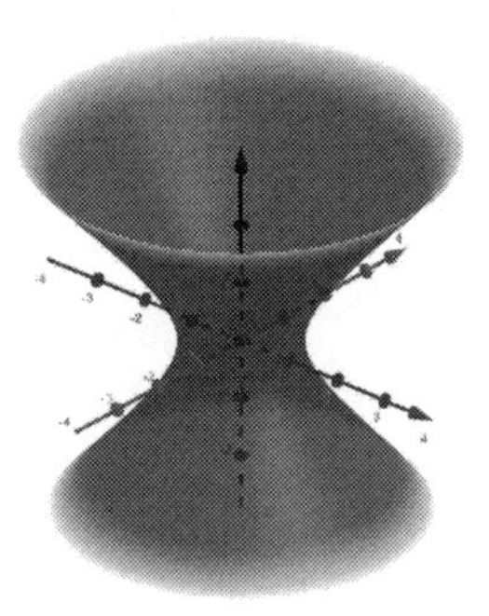

Made in the USA
Las Vegas, NV
16 May 2024